CHARLES DARWIN
INTERVIEWS AND RECOLLECTIONS

CHARLES DARWIN

Interviews and Recollections

Edited by

HAROLD OREL
University Professor Emeritus
English Department
University of Kansas

First published in Great Britain 2000 by
MACMILLAN PRESS LTD
Houndmills, Basingstoke, Hampshire RG21 6XS and London
Companies and representatives throughout the world

A catalogue record for this book is available from the British Library.

ISBN 0–333–72756–8

First published in the United States of America 2000 by
ST. MARTIN'S PRESS, INC.,
Scholarly and Reference Division,
175 Fifth Avenue, New York, N.Y. 10010

ISBN 0–312–22100–2

Library of Congress Cataloging-in-Publication Data
Charles Darwin : interviews and recollections / edited by Harold Orel.
p. cm.
Includes bibliographical references and index.
ISBN 0–312–22100–2 (cloth)
1. Darwin, Charles, 1809–1882. 2. Naturalists—England Biography.
I. Orel, Harold, 1926– .
QH31.D2C527 1999
576.8'092—dc21
[B]   99–18934
                                                              CIP

This book is printed on paper suitable for recycling and made from fully managed and
sustained forest sources.

10   9   8   7   6   5   4   3   2   1
09 08 07 06 05 04 03 02 01 00

Printed and bound in Great Britain by
Antony Rowe Ltd, Chippenham, Wiltshire

To Bill and Elaine House, with love

Contents

Each title is followed by a date within brackets, identifying the number of years covered by the selection or selections.

Introduction   ix
Note on the Text   xii

INTERVIEWS AND RECOLLECTIONS

Charles Darwin, 'Preliminary Notice', in
  Erasmus Darwin **[1794–6]**   1
Charles Darwin, 'An Autobiographical Fragment',
  in *More Letters of Charles Darwin* **[1813–20]**   6
Charles Darwin, 'The Early Years', in *The Autobiography
  of Charles Darwin, 1809–1882* **[1809–31]**   12
'The Walking Tour in North Wales', in *The Life
  and Letters of the Reverend Adam Sedgwick [I]*,
  edited by John Willis Clark and Thomas
  McKenny Hughes **[1831]**   55
Charles Darwin, 'Joining Captain Fitz-Roy on
  HMS Beagle', in *The Correspondence of
  Charles Darwin, 1821–1836* **[1831]**   59
Robert Fitz-Roy, 'Darwin Helps Save the Lifeboats
  of *HMS Beagle*', in *Narrative of the Surveying
  Voyages of His Majesty's Ships Adventure and
  Beagle* **[1831–2]**   78
Sir Bartholomew James Sulivan, 'Impressions of
  Charles Darwin' in *Life and Letters of the Late
  Admiral Sir Bartholomew James Sulivan* **[1831–6]**   83
'Darwin as a Welcome Guest', in *Harriet
  Martineau's Autobiography* **[1836–40s]**   87
'Owen Studies a Darwin Specimen', in *The Life of
  Richard Owen*, by Reverend Richard Owen **[1838]**   88
Charles Darwin, 'The Death of Anne Elizabeth Darwin',
  in *The Correspondence of Charles Darwin* **[1851]**   91

Alfred Russel Wallace, 'A Developing Friendship
with Darwin', *My Life / A Record of Events and
Opinions* **[1858–81]** 96

'Adam Sedgwick's Reaction to *The Origin of Species*',
in *The Life and Letters of the Reverend Adam Sedgwick*,
edited by John Willis Clark and Thomas
McKenny Hughes **[1859–65]** 118

'The Quarrel between Owen and Darwin', in
The Life of Richard Owen **[1859–82]** 128

'A Great Friendship', in *Life and Letters of
Sir Joseph Dalton Hooker* **[1839–63]** 133

'Sir Charles Lyell and *The Origin of Species*', in
*Life, Letters and Journals of Sir Charles Lyell,
Bart.* **[1838–64]** 154

'A Family Friend Comments on *The Origin of
Species*', in *Harriet Martineau's Letters to Fanny
Wedgwood* **[1860]** 160

'A Philosopher Comments on *The Origin of Species*',
in *An Autobiography* **[1859]** 164

John Stevens Henslow, 'Marriage Advice; Defending
Darwin', in *Darwin and Henslow / The Growth
of an Idea / Letters 1831–1860* **[1838, 1860]** 166

'The Great Debate', in *Life and Letters of Thomas
Henry Huxley* **[1860]** 171

'How Darwin Encouraged Galton', in *Memories
of My Life*, by Francis Galton **[1855, 1869]** 186

Francis Darwin, 'A Character Sketch by
Darwin's Son', in *The Life and Letters of
Charles Darwin* **[1860s–1882]** 192

'The Deaths of Erasmus and Charles Darwin', in
Life and Letters of Sir Joseph Dalton Hooker **[1881–2]** 214

Thomas Henry Huxley, 'Charles Darwin',
in *Nature* **[1882]** 218

Thomas Henry Huxley, 'The Darwin Memorial',
in *Darwiniana / Essays* **[1885]** 221

Index 225

Introduction

Two well-known images of Charles Darwin, each reinforcing the other, have influenced our understanding of this great Victorian figure. One image is that of a statue created in 1885 for a designated niche halfway up a major staircase in the Museum of Natural History, Kensington. The famous sculptor Sir Joseph Boehm did everything he could to emphasize Darwin's dignity and composure. Darwin, life-size, sits in a stylized chair, sombrely contemplating the spectacle of visitors to the Museum. One of his eyebrows is quizzically cocked, and both delicately tapered hands are folded together. Thomas Henry Huxley spoke on the occasion of its dedication, eloquently reminding his audience of Darwin's crucial role in reshaping several sciences, and, in a larger sense, the modern world.

The other image is that of a sepia-tinted photograph taken in 1868 by Julia Margaret Cameron. (Her *oeuvre* would later become one of the most important collections in the photographic archives of the National Portrait Gallery.) Darwin, along with his family, living for five weeks in a house owned by Julia Cameron (Freshwater, on the Isle of Wight), was in special need of a vacation. Recovering from an unusually intense work-period, Darwin, at the time, impressed William Allingham as 'yellow, sickly, very quiet', a description that implies Darwin was suffering from jaundice; and a few years later 'haggard' was added, this time in connection with another well-known photograph taken by Leonard Darwin in 1874. But the 'vacation' (a term that must be modified, since Darwin never wholly gave up his scientific interests on such expeditions) included visits from Joseph Dalton Hooker, Thomas Appleton, Henry Wadsworth Longfellow and Alfred Tennyson; walks along the beach; and quality time with his children.

Darwin liked Julia Cameron's photograph 'very much better than any other' which had been taken of him. Henrietta, his daughter, also called the profile-image 'excellent'.

Both the statue and the photograph show us a contemplative man. Darwin's earnestness of purpose, clinging not merely to his facial expression but to his very garments, was genuine; his expression had not been assumed for the benefit of the public. As a consequence, generations of readers and schoolchildren have been strongly influenced by an impression of Darwin as a stern, humourless *paterfamilias* and pedant, convinced that his theories of evolution and natural selection were right not merely for his own era but for centuries to come. A closer examination of the testimony proffered by those who knew Darwin personally will seriously modify this judgement, and to this end a volume in the Interviews and Recollections Series, published by Macmillan, has been prepared in the form of a composite biography, one which includes several important statements written by Darwin himself. He was a loving and (on numerous occasions) permissive father, remembered affectionately by his children. He knew well the limitations of the scientific research he had conducted, and the multiple lines of inquiry that had not achieved closure in his own work, as well as the limitations of other scientists. He was candid about these problems, and knew that the passage of time would seriously affect the validity of his work; he willingly confessed the presence of these problems in his publications; and if occasionally he expressed a strong dismay at the news that investigators of similar problems were about to pre-empt his findings by publication deadlines that he could not beat, his aim was to benefit from the recognition that he believed was due him for work that he had already completed rather than for research that was still on-going and inconclusive. His behaviour as a scientist in the developing fields of anthropology, biology, botany and geology constituted a role-model for his contemporaries, yet there was nothing priggish – and certainly very little that was affected by egotism – in either his public or his private life.

The facts having to do with Darwin the human being (as opposed to Darwin as the intrepid seeker after truth), recorded by his many friends and even some of his enemies, confirm the truth of such a characterization. Francis Darwin, his son, wrote a memoir that praised his father's work habits, sociability, literary preferences and writing style ('it is characteristic of himself in its simplicity, bordering on naivety, and in its absence of pretence'); he was in an ideal position to say so, and had achieved his own eminent position as a scientist when he recorded impressions of his father several years after his death. The reader, gathering impressions of Darwin's personal life, may want to judge Darwin the man by considering his notes on his grandfather and father, his memories of university days, and the actions he took in order to secure a berth aboard *HMS Beagle*. These autobiographical fragments (as well as the letters) are straightforward, and do not contradict the favourable evaluations of his character made by Sir Joseph Hooker, Sir Bartholomew James Sulivan, Robert Fitz-Roy, Alfred Russel Wallace, Adam Sedgwick and Thomas Henry Huxley.

All these witnesses, as well as several others who knew Darwin personally, are called upon for the making of this anthology. Taken together, their impressions illuminate a figure seen all too often only in silhouette. Darwin possessed an attractive, singular sense of his own character, one which remained largely undamaged despite countless fierce attacks against him made on religious or scientific grounds. If we come to understand the full dimensionality of Charles Darwin, we should be better able to appreciate the gigantic contribution he made to the multiple worlds of science and human behaviour.

Note on the Text

The text of the Readings has been standardized from the sometimes inconsistent spelling and punctuation of the original texts, to conform with modern British conventions: for example, single quotation marks, and the '-ize' spelling for verbs.

Charles Darwin, 'Preliminary Notice', in *Erasmus Darwin*, by Ernst Krause (New York: D. Appleton and Company, 1880), pp. 40–3, 84–5, 105–6 [1794–6]

My father spoke of Dr Darwin as having great powers of conversation. Lady Charleville, who had been accustomed to the most brilliant society in London, told him that Dr Darwin was one of the most agreeable men whom she had ever met. He himself used to say 'there were two sorts of agreeable persons in conversation parties – agreeable talkers and agreeable listeners'.

He stammered greatly, and it is surprising that this defect did not spoil his powers of conversation. A young man once asked him in, as he thought, an offensive manner, whether he did not find stammering very inconvenient. He answered, 'No, Sir, it gives me time for reflection, and saves me from asking impertinent questions.' Miss Seward speaks of him as being extremely sarcastic, but of this I can find no evidence in his letters or elsewhere. It is a pity that Dr Johnson in his visits to Lichfield rarely met Dr Darwin; but they seem to have disliked each other cordially, and to have felt that if they met they would have quarrelled like two dogs....

[Erasmus Darwin] possessed, according to my father, great facility in explaining any difficult subject; and he himself attributed this power to his habit of always talking about whatever he was studying, 'turning and moulding the subject according to the capacity of his hearers'. He compared himself to Gil Blas's uncle, who learned the grammar by teaching it to his nephew.[1]

1

When he wished to make himself disagreeable for any good cause, he was well able to do so. Lady ∗∗∗ married a widower, and became so jealous of his former wife that she cut and spoiled her picture, which hung up in one of the rooms. The husband, fearing that his young wife was becoming insane, was greatly alarmed, and sent for Dr Darwin. When he arrived he told her in the plainest manner many unpleasant truths, amongst others that the former wife was infinitely her superior in every respect, including beauty. The poor lady was astonished at being thus treated, and could never afterwards endure his name. He told the husband if she again behaved oddly, to hint that he would be sent for. The plan succeeded perfectly, and she ever afterwards restrained herself.

My father was much separated from Dr Darwin after early life, so that he remembered few of his remarks, but he used to quote one saying as very true: 'that the world was not governed by the clever men, but by the active and energetic.' He used also to quote another saying, that 'common sense would be improving, when men left off wearing as much flour on their heads as would make a pudding; when women left off wearing rings in their ears, like savages wear nose rings; and when fire-grates were no longer made of polished steel.'

Dr Darwin has been frequently called an atheist, whereas in every one of his works distinct expressions may be found showing that he fully believed in God as the Creator of the universe. For instance, in the 'Temple of Nature', published posthumously, he writes: 'Perhaps all the productions of nature are in their progress to greater perfection! an idea countenanced by modern discoveries and deductions concerning the progressive formation of the solid parts of the terraqueous globe, and consonant to the dignity of the creator of all things.' He concludes one chapter in 'Zoönomia' with the words of the Psalmist: '*The heavens declare the Glory of God, and the firmament sheweth his handiwork.*' ...

The third son, Robert Waring Darwin (my father, born 1766), did not inherit any aptitude for poetry or mechanics, nor did he possess, as I think, a scientific mind. He published, in vol. lxxvi of the 'Philosophical Transactions', a paper on Ocular Spectra, which Wheatstone[2] told me was a remarkable production for the period; but I believe that he was largely aided in writing it by his father. He was elected a Fellow of the Royal Society in 1788.[3] I cannot tell why my father's mind did not appear to me fitted for advancing science; for he was fond of theorizing, and was incomparably the most acute observer whom I ever knew. But his powers in this direction were exercised almost wholly in the practice of medicine, and in the observation of human character. He intuitively recognized the disposition or character, and even read the thoughts, of those with whom he came into contact with extraordinary acuteness. This skill partly accounts for his great success as a physician, for it impressed his patients with belief in him; and my father used to say that the art of gaining confidence was the chief element in a doctor's worldly success.

Erasmus brought him to Shrewsbury before he was twenty-one years old, and left him £20, saying, 'Let me know when you want more, and I will send it you.' His uncle, the rector of Elston, afterwards also sent him £20, and this was the sole pecuniary aid which he ever received. I have heard him say that his practice during the first year allowed him to keep two horses and a man-servant....

The 'Zoönomia' is largely devoted to medicine, and my father thought that it had much influenced medical practice in England; he was of course a partial, yet naturally a more observant judge than others on this point. The book when published was extensively read by the medical men of the day, and the author was highly esteemed by them as a practitioner. The following curious story, written down by his daughter, Violetta, in her old age,[4] shows his repute as a physician. A gentleman in the last stage of consumption came to Dr Darwin at Derby, and expressed himself to this effect: 'I am come from London to consult you, as the greatest physician in the world, to hear from you if there is any hope in my case; I know that my life hangs upon a thread, but while there is life there may be hope. It is of the utmost importance for me to settle my worldly affairs immediately; therefore I trust that you will not deceive me, but tell me without hesitation your candid opinion.' Dr Darwin felt his pulse, and minutely examined him, and said he was sorry to say there was no hope. After a pause of a few minutes the gentleman said: 'How long can I live?' The answer was: 'Perhaps a fortnight' The gentleman seized Dr Darwin's hand and said: 'Thank you, doctor, I thank you; my mind is satisfied; I now know there is no hope for me.' Dr Darwin then said: 'But as you come from London, why did you not consult Dr Warren, so celebrated a physician?' 'Alas! doctor, I am Dr Warren.' He died in a week or two afterwards.

NOTES

Erasmus Darwin (1731–1802), Charles Darwin's grandfather, enjoyed great fame during his lifetime. His *Botanic Garden* (1789–91) was widely read, even though its excessively mannered poetry always counted for much less than its enthusiastic advocacy of science. Other works include *Zoönomia, or the Laws of Organic Life* (1794–6), which anticipated some of Lamarck's views on generation (Charles would later speak of the 'erroneous grounds of opinion of Lamarck'); *A Plan for the Conduct of Female Education of Boarding Schools* (1797); *Phytologia, or the Philosophy of Agriculture and Gardening* (1800); and the posthumously published *The Temple of Nature* (1803).

Robert Waring Darwin, the third son of Erasmus Darwin by his first marriage, married a daughter of Josiah Wedgwood (1730–95), and in time became the father of Charles Darwin. Charles was the fifth child. His older brother, Erasmus Alvey Darwin (1804–81), differed strikingly from him, in that he was far more extrovert, his primary interests lay in literature, and he was a welcome member of many intersecting social circles. His death, though it had been anticipated, came as a severe blow to Charles.

Ernst Krause's biographical sketch of Erasmus Darwin appeared first in *Kosmos*, a well-known German scientific journal (February 1879). Charles Darwin prepared his 'memoir' (not a memoir based on personal knowledge, however, since Erasmus Darwin had died seven years before Charles was born) to accompany the English translation. In his Preliminary Notice Charles identified his sources, and indicated that he had undertaken serious research and had investigated all the available documents. Necessarily devoid of personal knowledge, Charles's essay is nevertheless a surprisingly objective consideration of an unconventional career, and must be considered a notable and pleasant family chronicle written in the spirit of his own *Autobiography*. The spirited defence that Charles mounts against the charge of atheism reminds us that the powerful threat of a similar charge against himself had led, two decades earlier, to some alterations of the text of *The Origin of Species*.

1.   *Gil Blas* (1715–35), by Alain René Le Sage (1668–1747).

2.   Charles Wheatstone (1802–75), inventor of the concertina and the stereoscope; a noted experimenter in the emerging sciences of acoustics, optics, electricity and telegraphy; and the Professor of Experimental Physics at King's College, London.

3.   The Darwin family, beginning with Erasmus Darwin, saw five successive generations nominated and elected to the Royal Academy.

4.   Violetta, the eldest daughter of Robert Waring Darwin by his second marriage, was the mother of Francis Galton, who became a distinguished anthropologist.

Charles Darwin, 'An Autobiographical Fragment', in *More Letters of Charles Darwin / A Record of his Work in a Series of Hitherto Unpublished Letters*, edited by Francis Darwin (New York: D. Appleton and Company, 1903), vol. I, pp. 1–5, 30 [1813–20]

My earliest recollection, the date of which I can approximately tell, and which must have been before I was four years old, was when sitting on Caroline's[1] knee in the drawing room, whilst she was cutting an orange for me, a cow ran by the window which made me jump, so that I received a bad cut, of which I bear the scar to this day. Of this scene I recollect the place where I sat and the cause of the fright, but not the cut itself, and I think my memory is real, and not as often happens in similar cases, [derived] from hearing the thing often repeated, [when] one obtains so vivid an image, that it cannot be separated from memory: because I clearly remember which way the cow ran, which would not probably have been told me. My memory here is an obscure picture, in which from not recollecting any pain I am scarcely conscious of its reference to myself.

1813.   When I was four years and a half old I went to the sea, and stayed there some weeks. I remember many things, but with the exception of the maidservants (and these are not individualized) I recollect none of my family who were there. I remember either myself or Catherine[2] being naughty, and being shut up in a room and trying to break the windows.

I have an obscure picture of a house before my eyes, and of a neighbouring small shop, where the owner gave me one fig, but which to my great joy turned out to be two: this fig was given me that the man might kiss the maidservant. I remember a common walk to a kind of well, on the road to which was a cottage shaded with damascene[3] trees, inhàbited by an old man, called a hermit, with white hair, who used to give us damascenes. I know not whether the damascenes, or the reverence and indistinct fear for this old man produced the greatest effect on my memory. I remember when going there crossing in the carriage a broad ford, and fear and astonishment of white foaming water has made a vivid impression. I think memory of events commences abruptly; that is, I remember these earliest things quite as clearly as others very much later in life, which were equally impressed on me. Some very early recollections are connected with fear at Parkfield[4] and with poor Betty Harvey. I remember with horror her story of people being pushed into the canal by the towing rope, by going the wrong side of the horse. I had the greatest horror of this story – keen instinct against death. Some other recollections are those of vanity – namely, thinking that people were admiring me, in one instance for perseverance and another for boldness in climbing a low tree, and what is odder, a consciousness, as if instinctive, that I was vain, and contempt of myself. My supposed admirer was old Peter Haile the bricklayer, and the tree the mountain ash on the lawn. All my recollections seem to be connected most closely with myself; now Catherine seems to recollect scenes where others were the chief actors. When my mother[5] died I was 8½ years old, and [Catherine] one year less, yet she remembers all particulars and events of each day whilst I scarcely recollect anything (and so with very many other cases) except being sent for the memory of going into her room, my father meeting me – crying afterwards. I recollect my mother's gown and scarcely anything of her appearance, except one or two walks with her. I have no distinct remembrance of any conversation, and those only of a very trivial nature. I remember her saying 'if she did ask me to do something,' which I said she had, 'it was solely for my good.'

Catherine remembers my mother crying, when she heard of my grandmother's death. Also when at Parkfield how Aunt Sarah and Aunt Kitty[6] used to receive her. Susan, like me, only remembers affairs personal. It is sufficiently odd this [difference] in subjects remembered. Catherine says she does not remember the impression made upon her by external things, as scenery, but for things which she reads she has an excellent memory, *i.e.*, for ideas. Now her sympathy being ideal, it is part of her character, and shows how easily her kind of memory was stamped, a vivid thought is repeated, a vivid impression forgotten.

I remember obscurely the illumination after the battle of Waterloo, and the Militia exercising about that period, in the field opposite our house.

1817.   At $8\frac{1}{2}$ years old I went to Mr Case's School.[7] I remember how very much I was afraid of meeting the dogs in Barker Street, and how at school I could not get up my courage to fight. I was very timid by nature. I remember I took great delight at school in fishing for newts in the quarry pool. I had thus young formed a strong taste for collecting, chiefly seals, franks, etc., but also pebbles and minerals – one which was given me by some boy decided this taste. I believe shortly after this, or before, I had smattered in botany, and certainly when at Mr Case's School I was very fond of gardening, and invented some great falsehoods about being able to colour crocuses as I liked. At this time I felt very strong friendship for some boys. It was soon after I began collecting stones, *i.e.*, when 9 or 10, that I distinctly recollect the desire I had of being able to know something about every pebble in front of the hall door – it was my earliest and only geological aspiration at that time. I was in those days a very great story-teller – for the pure pleasure of exciting attention and surprise. I stole fruit and hid it for these same motives, and injured trees by barking them for similar ends. I scarcely ever went out walking without saying I had seen a pheasant or some strange bird (natural history taste); these lies, when not detected, I presume, excited my attention, as I recollect them vividly, not connected with shame, though some I do, but as something which by having produced a great effect on my mind, gave pleasure like a tragedy. I recollect when I was at Mr Case's inventing a whole fabric to show how fond I was of speaking the *truth*! My invention is still so vivid in my mind, that I could almost fancy it was true, did not memory of former shame tell me it was false. I have no particularly happy or unhappy recollections of this time or earlier periods of my life. I remember well a walk I took with a boy named Ford across some fields to a farmhouse on the Church Stretton road. I do not remember any mental pursuits excepting those of collecting stones, etc., gardening, and about this time often going with my father in his carriage, telling him of my lessons, and seeing game and other wild birds, which was a great delight to me. I was born a naturalist.

When I was $9\frac{1}{2}$ years old (July 1818) I went with Erasmus to see Liverpool: it has left no impressions on my mind, except most trifling ones – fear of the coach upsetting, a good dinner, and an extremely vague memory of ships.

In Midsummer of this year I went to Dr Butler's School. I well recollect the first going there, which oddly enough I cannot of going to Mr Case's, the first school of all. I remember the year 1818 well, not from having first gone to a public school, but from writing those figures in my school book, accompanied with obscure thoughts, now fulfilled, whether I should recollect in future life that year.

In September (1818) I was ill with the scarlet fever. I well remember the wretched feeling of being delirious.

1819, July ($10\frac{1}{2}$ years old). Went to the sea at Plas Edwards[8] and stayed there three weeks, which now appears to me like three months. I remember a certain shady green road (where I saw a snake) and a waterfall, with a degree of pleasure, which must be connected with the pleasure from scenery, though not directly recognized as such. The sandy plain before the house has left a strong impression, which is obscurely connected with an indistinct remembrance of curious insects, probably a *Cimex* mottled with red, and

Zygaena, the burnet-moth. I was at that time very passionate (when I swore like a trooper) and quarrelsome. The former passion has I think nearly wholly but slowly died away. When journeying there by stage coach I remember a recruiting officer (I think I should know his face to this day) at tea time, asking the maid-servant for toasted bread and butter. I was convulsed with laughter and thought it the quaintest and wittiest speech that ever passed from the mouth of man. Such is wit at $10\frac{1}{2}$ years old. The memory now flashes across me of the pleasure I had in the evening on a blowy day walking along the beach by myself and seeing the gulls and cormorants wending their way home in a wild and irregular course. Such poetic pleasures, felt so keenly in after years, I should not have expected so early in life.

1820, July.   Went a riding tour (on old Dobbin) with Erasmus to Pistyll Rhiadr;[9] of this I recollect little, an indistinct picture of the fall, but I well remember my astonishment on hearing that fishes could jump up it.

•   •   •

[The following passage is taken from the MS copy of the *Autobiography*; it was not published in the *Life and Letters* which appeared in Mrs Darwin's lifetime. – F. D.]

You all know your mother, and what a good mother she has ever been to all of you. She has been my greatest blessing, and I can declare that in my whole life I have never heard her utter one word I would rather have been unsaid. She has never failed in kindest sympathy towards me, and has borne with the utmost patience my frequent complaints of ill-health and discomfort. I do not believe she has ever missed an opportunity of doing a kind action to any one near her. I marvel at my good fortune that she, so infinitely my superior in every single moral quality, consented to be my wife. She has been my wise adviser and cheerful comforter throughout life, which without her would have been during a very long period a miserable one from ill-health. She has earned the love of every soul near her.

NOTES

Francis Darwin has written that this autobiographical fragment, dated in 1838, was discovered 'in the process of removing the remainder of Mr Darwin's books and papers from Down'. Charles Darwin, born on 12 February 1809, here records a scattered number of recollections drawn from his first fifteen years, at the Mount; in his father's substantial red-brick house; and in Shrewsbury, where he attended two schools. He was the second son, and the fifth of six children. His short essay prefigured a more substantial effort in his *Autobiography*, written (for the most part) during a four-month period in mid-1876, but it contains some information not contained in the later work, and its value to readers interested in Darwin's life lies partly in the fact that Darwin was considerably younger (not yet thirty) when he took the trouble to record his impressions of childhood.

1.   Caroline Darwin (1800–88), his sister, was nine years older than he, and acted as a teacher and governess to him and to his younger sister.

2.   Emily Catherine Darwin (1810–66), his younger sister (by one year).

3.   Damson is derived from Damascene; the fruit was formerly known as a 'Damask prune'. [FD]

4.   Parkfield was the home of Mrs Josiah Wedgwood.

5.   Susannah Darwin (1765–1817), the daughter of Josiah Wedgwood, was Charles Darwin's mother. Her fragile health deteriorated, to the accompaniment of great suffering, before she died (most probably of peritonitis, for which nothing could be done). On the tombstone of Robert Waring Darwin, her husband, her name is written as 'Susan'.

6.   Sarah Elizabeth (1778–1856) and Catherine [Kitty] Wedgwood (1774–1823) were sisters.

7.   Charles Darwin went to this primary day-school, located in the High Street, in Shrewsbury (the county town of Shropshire) for only one year (1818–19). The Reverend George Case, minister at the Unitarian Chapel, was not a hard-driving converter of souls. Charles's mother, who helped to bring Case to Shrewsbury, attended chapel meetings; her husband liked to quote Erasmus Darwin (his father), to the effect that Unitarianism was merely a featherbed to catch a falling Christian. At the end of his first year Charles was sent to the public school, known as Shrewsbury School. Its headmaster was Samuel Butler, destined to become the grandfather of the author of *The Way of All Flesh*. (The novel satirizes him as Dr Skinner of Roughborough School.) Charles was a student in Shrewsbury between his ninth and sixteenth years, or 1818–25.

8.   At Towyn, on the Welsh coast.

9.   A stream that runs from Llyn Pen Rhiadr down the Blyfnant to the Dovey.

Charles Darwin, 'The Early Years', in *The Autobiography of Charles Darwin, 1809–1882*, edited by Nora Barlow (New York: W. W. Norton, 1958), pp. 21–71 [1809–31]

A German editor having written to me to ask for an account of the development of my mind and character with some sketch of my autobiography, I have thought that the attempt would amuse me, and might possibly interest my children or their children. I know that it would have interested me greatly to have read even so short and dull a sketch of the mind of my grandfather written by himself, and what he thought and did and how he worked. I have attempted to write the following account of myself, as if I were a dead man in another world looking back at my own life. Nor have I found this difficult, for life is nearly over with me. I have taken no pains about my style of writing.

I was born at Shrewsbury on February 12th, 1809. I have heard my Father say that he believed that persons with powerful minds generally had memories extending far back to a very early period of life. This is not my case for my earliest recollection goes back only to when I was a few months over four years old, when we went to near Abergele for sea-bathing, and I recollect some events and places there with some little distinctness.

My mother died in July 1817, when I was a little over eight years old, and it is odd that I can remember hardly anything about her except her death-bed, her black velvet gown, and her curiously constructed work-table. I believe that my forgetfulness is partly due to my sisters, owing to their great grief, never being able to speak about her or mention her name; and partly to her previous invalid state. In the spring of this same year I was sent to a day-school in Shrewsbury,[1] where I staid a year. Before going to school I was educated by my sister Caroline, but I doubt whether this plan answered. I have been told that I was much slower in learning than my younger sister Catherine, and I believe that I was in many ways a naughty boy. Caroline was extremely kind, clever and zealous; but she was too zealous in trying to improve me; for I clearly remember after this long interval of years, saying to myself when about to enter a room where she was – 'What will she blame me for now?' and I made myself dogged so as not to care what she might say.

By the time I went to this day-school my taste for natural history, and more especially for collecting, was well developed. I tried to make out the names of plants, and collected all sorts of things, shells, seals, franks, coins, and minerals. The passion for collecting, which leads a man to be a systematic naturalist, a virtuoso or a miser, was very strong in me, and was clearly innate, as none of my sisters or brother ever had this taste.

One little event during this year has fixed itself very firmly in my mind, and I hope that it has done so from my conscience having been afterwards sorely troubled by it; it is curious as showing that apparently I was interested at this early age in the variability of plants! I told another little boy (I believe it was Leighton,[2] who afterwards become a well-known Lichenologist and botanist) that I could produce variously coloured Polyanthuses and Primroses by watering them with certain coloured fluids, which was of course a monstrous fable, and had never been tried by me. I may here also confess that as a little boy I was much given to inventing deliberate falsehoods, and this was always done for the sake of causing excitement. For instance, I once gathered much valuable fruit from my Father's trees and hid them in the shrubbery, and then ran in breathless haste to spread the news that I had discovered a hoard of stolen fruit.[3]

About this time, or as I hope at a somewhat earlier age, I sometimes stole fruit for the sake of eating it; and one of my schemes was ingenious. The kitchen garden was kept locked in the evening, and was surrounded by a high wall, but by the aid of neighbouring trees I could easily get on the coping. I then fixed a long stick into the hole at the bottom of a rather large flower-pot, and by dragging this upwards pulled off peaches and plums, which fell into the pot and the prizes were thus secured. When a very little boy I remember stealing apples from the orchard, for the sake of giving them away to some boys and young men who lived in a cottage not far off, but before I gave them the fruit I showed off how quickly I could run and it is wonderful that I did not perceive that the surprise and admiration which they expressed at my powers of running, was given for the sake of the apples. But I well remember that I was delighted at them declaring that they had never seen a boy run so fast!

I remember clearly only one other incident during the years whilst at Mr Case's daily school – namely, the burial of a dragoon-soldier; and it is surprising how clearly I can still see the horse with the man's empty boots and carbine suspended to the saddle, and the firing over the grave. This scene deeply stirred whatever poetic fancy there was in me.

In the summer of 1818 I went to Dr Butler's great school in Shrewsbury, and remained there for seven years till Mid-summer 1825, when I was sixteen years old. I boarded at this school, so that I had the great advantage of living the life of a true school-boy; but as the distance was hardly more than a mile to my home, I very often ran there in the longer intervals between the callings over and before lock-ing up at night. This I think was in many ways advantage-ous to me by keeping up home affections and interests. I remember in the early part of my school life that I often had to run very quickly to be in time, and from being a fleet run-ner was generally successful; but when in doubt I prayed earnestly to God to help me, and I well remember that I attributed my success to the prayers and not to my quick running, and marvelled how generally I was aided.

I have heard my father and elder sisters say that I had, as a very young boy, a strong taste for long solitary walks; but what I thought about I know not. I often became quite absorbed, and once, whilst returning to school on the sum-mit of the old fortifications round Shrewsbury, which had been converted into a public foot-path with no parapet on one side, I walked off and fell to the ground, but the height was only seven or eight feet. Nevertheless the number of thoughts which passed through my mind during this very short, but sudden and wholly unexpected fall, was aston-ishing, and seem hardly compatible with what physiolo-gists have, I believe, proved about each thought requiring quite an appreciable amount of time.

I must have been a very simple little fellow when I first went to the school. A boy of the name of Garnett took me into a cake-shop one day, and bought some cakes for which he did not pay, as the shopman trusted him. When we came out I asked him why he did not pay for them, and he instantly answered, 'Why, do you not know that my uncle left a great sum of money to the Town on condition that every tradesman should give whatever was wanted without payment to anyone who wore his old hat and moved it in a particular manner;' and he then showed me how it was moved. He then went into another shop where he was trusted, and asked for some small article, moving his hat in the proper manner, and of course obtained it without payment. When we came out he said, 'Now if you like to go by yourself into that cake-shop (how well I remember its exact position), I will lend you my hat, and you can get whatever you like if you move the hat on your head properly.' I gladly accepted the generous offer, and went in and asked for some cakes, moved the old hat, and was walking out of the shop, when the shopman made a rush at me, so I dropped the cakes and ran away for dear life, and was astonished by being greeted with shouts of laughter by my false friend Garnett.

I can say in my own favour that I was as a boy humane, but I owed this entirely to the instruction and example of my sisters. I doubt indeed whether humanity is a natural or innate quality. I was very fond of collecting eggs, but I never took more than a single egg out of a bird's nest, except on one single occasion, when I took all, not for their value, but from a sort of bravado.

I had a strong taste for angling, and would sit for any number of hours on the bank of a river or pond watching the float; when at Maer[4] I was told that I could kill the worms with salt and water, and from that day I never spitted a living worm, though at the expense, probably, of some loss of success.

Once as a very little boy, whilst at the day-school, or before that time, I acted cruelly, for I beat a puppy I believe, simply from enjoying the sense of power; but the beating could not have been severe, for the puppy did not howl, of which I feel sure as the spot was near to the house. This act lay heavily on my conscience, as is shown by my remembering the exact spot where the crime was committed. It probably lay all the heavier from my love of dogs being then, and for a long time afterwards, a passion. Dogs seemed to know this, for I was an adept in robbing their love from their masters.

Nothing could have been worse for the development of my mind than Dr Butler's school, as it was strictly classical, nothing else being taught except a little ancient geography and history. The school as a means of education to me was simply a blank. During my whole life I have been singularly incapable of mastering any language. Especial attention was paid to verse-making, and this I could never do well. I had many friends, and got together a grand collection of old verses, which by patching together, sometimes aided by other boys, I could work into any subject. Much attention was paid to learning by heart the lessons of the previous day; this I could effect with great facility learning forty or fifty lines of Virgil or Homer, whilst I was in morning chapel; but this exercise was utterly useless, for every verse was forgotten in forty-eight hours. I was not idle, and with the exception of versification, generally worked conscientiously at my classics, not using cribs. The sole pleasure I ever received from such studies, was from some of the odes of Horace, which I admired greatly. When I left the school I was for my age neither high nor low in it; and I believe that I was considered by all my masters and by my Father as a very ordinary boy, rather below the common standard in intellect. To my deep mortification my father once said to me, 'You care for nothing but shooting, dogs, and rat-catching, and you will be a disgrace to yourself and all your family.' But my father, who was the kindest man I ever knew, and whose memory I love with all my heart, must have been angry and somewhat unjust when he used such words.

I may here add a few pages about my Father, who was in many ways a remarkable man.[5]

He was about 6 feet 2 inches in height, with broad shoulders, and very corpulent, so that he was the largest man whom I ever saw. When he last weighed himself, he was 24 stone, but afterwards increased much in weight. His chief mental characteristics were his powers of observation and his sympathy, neither of which have I ever seen exceeded or even equalled. His sympathy was not only with the distresses of others, but in a greater degree with the pleasures of all around him. This led him to be always scheming to give pleasure to others, and, though hating extravagance, to perform many generous actions. For instance, Mr B——, a small manufacturer in Shrewsbury, came to him one day, and said he should be bankrupt unless he could at once borrow £10,000, but that he was unable to give any legal security. My father heard his reasons for believing that he could ultimately repay the money, and from my Father's intuitive perception of character felt sure that he was to be trusted. So he advanced this sum, which was a very large one for him while young, and was after a time repaid.

I suppose that it was his sympathy which gave him unbounded power of winning confidence, and as a consequence made him highly successful as a physician. He began to practise before he was twenty-one years old, and his fees during the first year paid for the keep of two horses and a servant. On the following year his practice was larger, and so continued for above sixty years, when he ceased to attend on any one. His great success as a doctor was the more remarkable, as he told me that he at first hated his profession so much that if he had been sure of the smallest pittance, or if his father had given him any choice, nothing should have induced him to follow it. To the end of his life, the thought of an operation almost sickened him, and he could scarcely endure to see a person bled – a horror which he has transmitted to me – and I remember the horror which I felt as a schoolboy in reading about Pliny (I think) bleeding to death in a warm bath. My Father told me two odd stories about bleeding: one was that as a very young man he became a Freemason. A friend of his who was a Freemason and who pretended not to know about his strong feeling with respect to blood, remarked casually to him as they walked to the meeting, 'I suppose that you do not care about losing a few drops of blood?' It seems that when he was received as a member, his eyes were bandaged and his coat-sleeves turned up. Whether any such ceremony is now performed I know not, but my Father mentioned the case as an excellent instance of the power of imagination, for he distinctly felt the blood trickling down his arm, and could hardly believe his own eyes, when he afterwards could not find the smallest prick on his arm.

A great slaughtering butcher from London once consulted my grandfather, when another man very ill was brought in, and my grandfather wished to have him instantly bled by the accompanying apothecary. The butcher was asked to hold the patient's arm, but he made some excuse and left the room. Afterwards he explained to my grandfather that although he believed that he had killed with his own hands more animals than any other man in London, yet absurd as it might seem he assuredly should have fainted if he had seen the patient bled.

Owing to my father's power of winning confidence, many patients, especially ladies, consulted him when suffering from any misery, as a sort of Father-Confessor. He told me that they always began by complaining in a vague manner about their health, and by practice he soon guessed what was really the matter. He then suggested that they had been suffering in their minds, and now they would pour out their troubles, and he heard nothing more about the body. Family quarrels were a common subject. When gentlemen complained to him about their wives, and the quarrel seemed serious, my Father advised them to act in the following manner; and his advice always succeeded if the gentleman followed it to the letter, which was not always the case. The husband was to say to the wife that he was very sorry that they could not live happily together, – that he felt sure that she would be happier if separated from him – that he did not blame her in the least (this was the point on which the man oftenest failed) – that he would not blame her to any of her relations or friends and lastly that he would settle on her as large a provision as he could afford. She was then asked to deliberate on this proposal. As no fault had been found, her temper was unruffled, and she soon felt what an awkward position she would be in, with no accusation to rebut, and with her husband and not herself proposing a separation. Invariably the lady begged her husband not to think of separation, and usually behaved much better ever afterwards.

Owing to my father's skill in winning confidence he received many strange confessions of misery and guilt. He often remarked how many miserable wives he had known. In several instances husbands and wives had gone on pretty well together for between twenty and thirty years, and then hated each other bitterly: this he attributed to their having lost a common bond in their young children having grown up.

But the most remarkable power which my father possessed was that of reading the characters, and even the thoughts of those whom he saw even for a short time. We had many instances of this power, some of which seemed almost supernatural. It saved my father from ever making (with one exception, and the character of this man was soon discovered) an unworthy friend. A strange clergyman came to Shrewsbury, and seemed to be a rich man; everybody called on him, and he was invited to many houses. My father called, and on his return home told my sisters on no account to invite him or his family to our house; for he felt sure that the man was not to be trusted. After a few months he suddenly bolted, being heavily in debt, and was found out to be little better than an habitual swindler. Here is a case of trustfulness which not many men would have ventured on. An Irish gentleman, a complete stranger, called on my father one day, and said that he had lost his purse, and that it would be a serious inconvenience to him to wait in Shrewsbury until he could receive a remittance from Ireland. He then asked my father to lend him £20, which was immediately done, as my father felt certain that the story was a true one. As soon as a letter could arrive from Ireland, one came with the most profuse thanks, and enclosing, as he said, a £20 Bank of England note; but no note was enclosed. I asked my father whether this did not stagger him, but he answered 'not in the least.' On the next day another letter came with many apologies for having forgotten (like a true Irishman) to put the note into his letter of the day before.

A connection[6] of my Father's consulted him about his son who was strangely idle and would settle to no work. My Father said 'I believe that the foolish young man thinks that I shall bequeath him a large sum of money. Tell him that I have declared to you that I shall not leave him a penny.' The Father of the youth owned with shame that this preposterous idea had taken possession of his son's mind; and he asked my Father how he could possibly have discovered it, but my Father said he did not in the least know.

The Earl of —— brought his nephew, who was insane but quite gentle, to my father; and the young man's insanity led him to accuse himself of all the crimes under heaven. When my Father afterwards talked about the case with the uncle, he said, 'I am sure that your nephew is really guilty of...a heinous crime.' Whereupon the Earl of —— exclaimed, 'Good God, Dr Darwin, who told you; we thought that no human being knew the fact except ourselves!' My Father told me the story many years after the event, and I asked him how he distinguished the true from the false self-accusations; and it was very characteristic of my Father that he said he could not explain how it was.

The following story shows what good guesses my Father could make. Lord Sherburn,[7] afterwards the first Marquis of Lansdowne, was famous (as Macaulay somewhere remarks) for his knowledge of the affairs of Europe, on which he greatly prided himself. He consulted my Father medically, and afterwards harangued him on the state of Holland. My father had studied medicine at Leyden, and one day went a long walk into the country with a friend, who took him to the house of a clergyman (we will say the Rev. Mr A——, for I have forgotten his name), who had married an

Englishwoman. My father was very hungry, and there was little for luncheon except cheese, which he could never eat. The old lady was surprised and grieved at this, and assured my father that it was an excellent cheese, and had been sent her from Bowood, the seat of Lord Sherburn. My father wondered why a cheese should be sent her from Bowood, but thought nothing more about it until it flashed across his mind many years afterwards, whilst Lord Sherburn was talking about Holland. So he answered, 'I should think from what I saw of the Rev. Mr A——, that he was a very able man and well acquainted with the state of Holland.' My father saw that the Earl, who immediately changed the conversation, was much startled. On the next morning my father received a note from the Earl, saying that he had delayed starting on his journey, and wished particularly to see my father. When he called, the Earl said, 'Dr Darwin, it is of the utmost importance to me and to the Rev. Mr A—— to learn how you have discovered that he is the source of my information about Holland.' So my father had to explain the state of the case, and he supposed that Lord Sherburn was much struck with his diplomatic skill in guessing, for during many years afterwards he received many kind messages from him through various friends. I think that he must have told the story to his children; for Sir C. Lyell asked me many years ago why the Marquis of Lansdowne (the son or grandson of the first marquis) felt so much interest about me, whom he had never seen, and my family. When forty new members (the forty thieves as they were then called) were added to the Athenaeum Club, there was much canvassing to be one of them; and without my having asked any one, Lord Lansdowne proposed me and got me elected. If I am right in my supposition, it was a queer concatenation of events that my father not eating cheese half-a-century before in Holland led to my election as a member of the Athenaeum.

Early in life my father occasionally wrote down a short account of some curious event and conversation, which are enclosed in a separate envelope.

The sharpness of his observation led him to predict with remarkable skill the course of any illness, and he suggested endless small details of relief. I was told that a young Doctor in Shrewsbury, who disliked my father, used to say that he was wholly unscientific, but owned that his power of predicting the end of an illness was unparalleled. Formerly when he thought that I should be a doctor, he talked much to me about his patients. In the old days the practice of bleeding largely was universal, but my father maintained that far more evil was thus caused than good done; and he advised me if ever I was myself ill not to allow any doctor to take from me more than an extremely small quantity of blood. Long before typhoid fever was recognized as distinct, my father told me that two utterly distinct kinds of illness were confounded under the name of typhus fever. He was vehement against drinking, and was convinced of both the direct and inherited evil effects of alcohol when habitually taken even in moderate quantity in a very large majority of cases.[8] But he admitted and advanced instances of certain persons, who could drink largely during their whole lives without apparently suffering any evil effects; and he believed that he could often beforehand tell who would thus not suffer. He himself never drank a drop of any alcoholic fluid. This remark reminds me of a case showing how a witness under the most favourable circumstances may be wholly mistaken. A gentleman-farmer was strongly urged by my father not to drink, and was encouraged by being told that he himself never touched any spirituous liquor. Whereupon the gentleman said, 'Come, come, Doctor, that won't do – though it is very kind of you to say so for my sake – for I know that you take a very large glass of hot gin and water every evening after your dinner.'[9] So my father asked him how he knew this. The man answered, 'My cook was your kitchen-maid for two or three years, and she saw the butler every day prepare and take to you the gin and water.' The explanation was that my father had the odd habit of drinking hot water in a very tall and large glass after his dinner; and the butler used first to put some cold water in the glass, which the girl mistook for gin, and then filled it up with boiling water from the kitchen boiler.

My father used to tell me many little things which he had found useful in his medical practice. Thus ladies often cried much while telling him their troubles, and thus caused much loss of his precious time. He soon found that begging them to command and restrain themselves, always made them weep the more, so that afterwards he always encouraged them to go on crying, saying that this would relieve them more than anything else, with the invariable result that they soon ceased to cry, and he could hear what they had to say and give his advice. When patients who were very ill, craved for some strange and unnatural food, my father asked them what had put such an idea into their heads: if they answered that they did not know, he would allow them to try the food, and often with success, as he trusted to their having a kind of instinctive desire; but if they answered that they had heard that the food in question had done good to someone else, he firmly refused his assent.

He gave one day an odd little specimen of human nature. When a very young man he was called in to consult with the family physician in the case of a gentleman of much distinction in Shropshire. The old doctor told the wife that the illness was of such a nature that it must end fatally. My father took a different view and maintained that the gentleman would recover: he was proved quite wrong in all respects, (I think by autopsy) and he owned his error. He was then convinced that he should never again be consulted by this family; but after a few months the widow sent for him, having dismissed the old family doctor. My father was so much surprised at this, that he asked a friend of the widow to find out why he was again consulted. The widow answered her friend, that 'she would never again see that odious old doctor who said from the first that her husband would die, while Dr Darwin always maintained that he would recover!' In another case my father told a lady that her husband would certainly die. Some months afterwards he saw the widow who was a very sensible woman, and she said, 'You are a very young man, and allow me to advise you always to give, as long as you possibly can, hope to any near relation nursing a patient. You made me despair, and from that moment I lost strength.' My father said that he had often since seen the paramount importance, for the sake of the patient, of keeping up the hope and with it the strength of the nurse in charge. This he sometimes found it difficult to do compatibly with truth. One old gentleman, however, Mr Pemberton, caused him no such perplexity. He was sent for by Mr Pemberton, who said, 'From all that I have seen and heard of you I believe you are the sort of man who will speak the truth, and if I ask you will tell me when I am dying. Now I much desire that you should attend me, if you will promise, whatever I may say, always to declare that I am not going to die.' My father acquiesced on this understanding that his words should in fact have no meaning.

My father possessed an extraordinary memory, especially for dates, so that he knew, when he was very old the day of the birth, marriage, and death of a multitude of persons in Shropshire; and he once told me that this power annoyed him; for if he once heard a date he could not forget it; and thus the deaths of many friends were often recalled to his mind. Owing to his strong memory he knew an extraordinary number of curious stories, which he liked to tell, as he was a great talker. He was generally in high spirits, and laughed and joked with every one – often with his servants – with the utmost freedom; yet he had the art of making every one obey him to the letter. Many persons were much afraid of him. I remember my father telling us one day with a laugh, that several persons had asked him whether Miss Piggott (a grand old lady in Shropshire), had called on him, so that at last he enquired why they asked him; and was told that Miss Piggott, whom my father had somehow mortally offended, was telling everybody that she would call and tell 'that fat old doctor very plainly what she thought of him'. She had already called, but her courage had failed, and no one could have been more courteous and friendly.

As a boy, I went to stay at the house of Major B——, whose wife was insane; and the poor creature, as soon as she saw me, was in the most abject state of terror that I ever saw, weeping bitterly and asking me over and over again, 'Is your father coming?' but was soon pacified. On my return home, I asked my father why she was so frightened, and he answered he [was] very glad to hear it, as he had frightened her on purpose, feeling sure that she could be kept in safety and much happier without any restraint, if her husband could influence her, whenever she became at all violent, by proposing to send for Dr Darwin; and these words succeeded perfectly during the rest of her long life.

My father was very sensitive so that many small events annoyed or pained him much. I once asked him, when he was old and could not walk, why he did not drive out for exercise; and he answered, 'Every road out of Shrewsbury is associated in my mind with some painful event.' Yet he was generally in high spirits. He was easily made very angry, but as his kindness was unbounded, he was widely and deeply loved.

He was a cautious and good man of business, so that he hardly ever lost money by any investment, and left to his children a very large property. I remember a story, showing how easily utterly false beliefs originate and spread. Mr E——, a squire of one of the oldest families in Shropshire, and head partner in a Bank, committed suicide. My father was sent for as a matter of form, and found him dead. I may mention by the way, to show how matters were managed in those old days, that because Mr E—— was a rather great man and universally respected, no inquest was held over his body. My father, in returning home, thought it proper to call at the Bank (where he had an account) to tell the managing partner of the event, as it was not improbable it would cause a run on the bank. Well the story was spread far and wide, that my father went into the bank, drew out all his money, left the bank, came back again, and said, 'I may just tell you that Mr E—— has killed himself,' and then departed. It seems that it was then a common belief that money withdrawn from a bank was not safe, until the person had passed out through the door of the bank. My father did not hear this story till some little time afterwards, when the managing partner said that he had departed from his invariable rule of never allowing any one to see the account of another man, by having shown the ledger with my father's account to several persons, as this proved that my father had not drawn out a penny on that day. It would have been dishonourable in my father to have used his professional knowledge for his private advantage. Nevertheless the supposed act was greatly admired by some persons; and many years afterwards, a gentleman remarked, 'Ah, Doctor, what a splendid man of business you were in so cleverly getting all your money safe out of that bank.'

My father's mind was not scientific, and he did not try to generalize his knowledge under general laws; yet he formed a theory for almost everything which occurred. I do not think that I gained much from him intellectually; but his example ought to have been of much moral service to all his children. One of his golden rules (a hard one to follow) was, 'Never become the friend of any one whom you cannot respect.'

With respect to my Father's father, the author of the *Botanic Garden* etc., I have put together all the facts which I could collect in his published *Life*.[10]

Having said this much about my Father, I will add a few words about my brother and sisters.

My brother Erasmus possessed a remarkably clear mind, with extensive and diversified tastes and knowledge in literature, art, and even in science. For a short time he collected and dried plants, and during a somewhat longer time experimented in chemistry. He was extremely agreeable, and his wit often reminded me of that in the letters and works of Charles Lamb. He was very kind-hearted; but his health from his boyhood had been weak, and as a consequence he failed in energy. His spirits were not high, sometimes low, more especially during early and middle manhood. He read much, even whilst a boy, and at school encouraged me to read, lending me books. Our minds and tastes were, however, so different that I do not think that I owe much to him intellectually – nor to my four sisters, who possessed very different characters, and some of them had strongly marked characters. All were extremely kind and affectionate towards me during their whole lives. I am inclined to agree with Francis Galton in believing that education and environment produce only a small effect on the mind of any one, and that most of our qualities are innate.

The above sketch of my brother's character was written before that which was published in Carlyle's Remembrances,[11] and which appears to me to have little truth and no merit.

Looking back as well as I can at my character during my school life, the only qualities which at this period promised well for the future, were, that I had strong and diversified tastes, much zeal for whatever interested me, and a keen pleasure in understanding any complex subject or thing. I was taught Euclid by a private tutor, and I distinctly remember the intense satisfaction which the clear geometrical proofs gave me. I remember with equal distinctness the delight which my uncle[12] gave me (the father of Francis Galton) by explaining the principle of the vernier of a barometer. With respect to diversified tastes, independently of science, I was fond of reading various books, and I used to sit for hours reading the historical plays of Shakespeare, generally in an old window in the thick walls of the school. I read also other poetry, such as the recently published poems of Byron, Scott, and Thomson's *Seasons*. I mention this because later in life I wholly lost, to my great regret, all pleasure from poetry of any kind, including Shakespeare. In connection with pleasure from poetry I may add that in 1822 a vivid delight in scenery was first awakened in my mind, during a riding tour on the borders of Wales, and which has lasted longer than any other aesthetic pleasure.

Early in my school-days a boy had a copy of the *Wonders of the World*,[13] which I often read and disputed with other boys about the veracity of some of the statements; and I believe this book first gave me a wish to travel in remote countries, which was ultimately fulfilled by the voyage of the *Beagle*. In the latter part of my school life I became passionately fond of shooting, and I do not believe that anyone could have shown more zeal for the most holy cause than I did for shooting birds. How well I remember killing my first snipe, and my excitement was so great that I had much difficulty in reloading my gun from the trembling of my hands. This taste long continued and I became a very good shot. When at Cambridge I used to practise throwing up my gun to my shoulder before a looking-glass to see that I threw it up straight. Another and better plan was to get a friend to wave about a lighted candle, and then to fire at it with a cap on the nipple, and if the aim was accurate the little puff of air would blow out the candle. The explosion of the cap caused a sharp crack, and I was told that the Tutor of the College remarked, 'What an extraordinary thing it is, Mr Darwin seems to spend hours in cracking a horse-whip in his room, for I often hear the crack when I pass under his windows.'

I had many friends amongst the schoolboys, whom I loved dearly, and I think that my disposition was then very affectionate. Some of these boys were rather clever, but I may add on the principle of 'noscitur a socio' that not one of them ever became in the least distinguished.

With respect to science, I continued collecting minerals with much zeal, but quite unscientifically – all that I cared for was a new *named* mineral, and I hardly attempted to classify them. I must have observed insects with some little care, for when ten years old (1819) I went for three weeks to Plas Edwards on the sea-coast in Wales, I was very much interested and surprised at seeing a large black and scarlet Hemipterous insect, many moths (Zygaena) and a Cicindela, which are not found in Shropshire. I almost made up my mind to begin collecting all the insects which I could find dead, for on consulting my sister, I concluded that it was not right to kill insects for the sake of making a collection.

From reading White's *Selborne*[14] I took much pleasure in watching the habits of birds, and even made notes on the subject. In my simplicity I remember wondering why every gentleman did not become an ornithologist.

Towards the close of my school life, my brother worked hard at chemistry and made a fair laboratory with proper apparatus in the tool-house in the garden, and I was allowed to aid him as a servant in most of his experiments. He made all the gases and many compounds, and I read with care several books on chemistry, such as Henry and Parkes' *Chemical Catechism*.[15] The subject interested me greatly, and we often used to go on working till rather late at night. This was the best part of my education at school, for it showed me practically the meaning of experimental science. The fact that we worked at chemistry somehow got known at school, and as it was an unprecedented fact, I was nicknamed 'Gas'. I was also once publicly rebuked by the head-master, Dr Butler, for thus wasting my time over such useless subjects; and he called me very unjustly a 'poco curante',[16] and as I did not understand what he meant it seemed to me a fearful reproach.

As I was doing no good at school, my father wisely took me away at a rather earlier age than usual, and sent me (October 1825) to Edinburgh University[17] with my brother, where I stayed for two years or sessions. My brother was completing his medical studies, though I do not believe he ever really intended to practise, and I was sent there to commence them. But soon after this period I became convinced from various small circumstances that my father would leave me property enough to subsist on with some comfort, though I never imagined that I should be so rich a man as I am; but my belief was sufficient to check any strenuous effort to learn medicine.

The instruction at Edinburgh was altogether by Lectures, and these were intolerably dull, with the exception of those on chemistry by Hope;[18] but to my mind there are no advantages and many disadvantages in lectures compared with reading. Dr Duncan's[19] lectures on Materia Medical at

8 o'clock on a winter's morning are something fearful to
remember. Dr Munro[20] made his lectures on human anatomy as dull, as he was himself, and the subject disgusted
me. It has proved one of the greatest evils in my life that
I was not urged to practise dissection, for I should soon
have got over my disgust; and the practice would have
been invaluable for all my future work. This has been an
irremediable evil, as well as my incapacity to draw. I also
attended regularly the clinical wards in the Hospital. Some
of the cases distressed me a good deal, and I still have vivid
pictures before me of some of them; but I was not so foolish
as to allow this to lessen my attendance. I cannot understand why this part of my medical course did not interest
me in a greater degree; for during the summer before coming to Edinburgh I began attending some of the poor
people, chiefly children and women in Shrewsbury: I wrote
down as full an account as I could of the cases with all the
symptoms, and read them aloud to my father, who suggested
further enquiries, and advised me what medicines to give,
which I made up myself. At one time I had at least a dozen
patients, and I felt a keen interest in the work.[21] My father,
who was by far the best judge of character whom I ever
knew, declared that I should make a successful physician,
– meaning by this, one who got many patients. He maintained that the chief element of success was exciting confidence; but what he saw in me which convinced him that I
should create confidence I know not. I also attended on two
occasions the operating theatre in the hospital at Edinburgh, and saw two very bad operations, one on a child, but
I rushed away before they were completed. Nor did I ever
attend again, for hardly any inducement would have been
strong enough to make me do so; this being long before the
blessed days of chloroform. The two cases fairly haunted
me for many a long year.

My Brother staid only one year at the University, so that
during the second year I was left to my own resources; and
this was an advantage, for I became well acquainted with
several young men fond of natural science. One of these

was Ainsworth,[22] who afterwards published his travels in
Assyria: he was a Wernerian[23] geologist and knew a little
about many subjects, but was superficial and very glib with
his tongue. Dr Coldstream[24] was a very different young man,
prim, formal, highly religious and most kind-hearted: he
afterwards published some good zoological articles. A third
young man was Hardie, who would I think have made
a good botanist, but died early in India.[25] Lastly, Dr Grant,[26]
my senior by several years, but how I became acquainted
with him I cannot remember: he published some first-rate
zoological papers, but after coming to London as Professor
in University College, he did nothing more in science –
a fact which has always been inexplicable to me. I knew
him well; he was dry and formal in manner, but with
much enthusiasm beneath this outer crust. He one day,
when we were walking together burst forth in high admiration of Lamarck and his views on evolution. I listened in
silent astonishment, and as far as I can judge, without
any effect on my mind. I had previously read the *Zoönomia*
of my grandfather, in which similar views are maintained,
but without producing any effect on me. Nevertheless it is
probable that the hearing rather early in life such views
maintained and praised may have favoured my upholding
them under a different form in my *Origin of Species*. At this
time I admired greatly the *Zoönomia*; but on reading it a
second time after an interval of ten or fifteen years, I was
much disappointed, the proportion of speculation being so
large to the facts given.

Drs Grant and Coldstream attended much to marine
Zoology, and I often accompanied the former to collect
animals in the tidal pools, which I dissected as well as I
could. I also became friends with some of the Newhaven
fishermen, and sometimes accompanied them when they
trawled for oysters, and thus got many specimens. But
from not having had any regular practice in dissection, and
from possessing only a wretched microscope my attempts
were very poor. Nevertheless I made one interesting little
discovery, and read about the beginning of the year 1826,

a short paper on the subject before the Plinian Socy.[27] This was that the so-called ova of Flustra had the power of independent movement by means of cilia, and were in fact larvae. In another short paper I showed that little globular bodies which had been supposed to be the young state of *Fucus loreus* were the egg-cases of the worm-like *Pontobdella muricata*.

The Plinian Society was encouraged and I believe founded by Professor Jameson:[28] it consisted of students and met in an underground room in the University for the sake of reading papers on natural science and discussing them. I used regularly to attend and the meetings had a good effect on me in stimulating my zeal and giving me new congenial acquaintances. One evening a poor young man got up and after stammering for a prodigious length of time, blushing crimson, he at last slowly got out the words, 'Mr President, I have forgotten what I was going to say.' The poor fellow looked quite overwhelmed, and all the members were so surprised that no one could think of a word to say to cover his confusion. The papers which were read to our little society were not printed, so that I had not the satisfaction of seeing my paper in print; but I believe Dr Grant noticed my small discovery in his excellent memoir on Flustra.

I was also a member of the Royal Medical Society, and attended pretty regularly, but as the subjects were exclusively medical I did not much care about them. Much rubbish was talked there, but there were some good speakers, of whom the best was the present Sir J. Kay-Shuttleworth.[29] Dr Grant took me occasionally to the meetings of the Wernerian Society, where various papers on natural history were read, discussed, and afterwards published in the Transactions. I heard Audubon[30] deliver there some interesting discourses on the habits of N. American birds, sneering somewhat unjustly at Waterton.[31] By the way, a negro lived in Edinburgh, who had travelled with Waterton and gained his livelihood by stuffing birds, which he did excellently; he gave me lessons for payment, and I used often to sit with him, for he was a very pleasant and intelligent man.

Mr Leonard Horner[32] also took me once to a meeting of the Royal Society of Edinburgh, where I saw Sir Walter Scott in the chair as President, and he apologized to the meeting as not feeling fitted for such a position. I looked at him and at the whole scene with some awe and reverence; and I think it was owing to this visit during my youth and to my having attended the Royal Medical Society, that I felt the honour of being elected a few years ago an honorary member of both these Societies, more than any other similar honour. If I had been told at that time that I should one day have been thus honoured, I declare that I should have thought it as ridiculous and improbable, as if I had been told that I should be elected King of England.

During my second year in Edinburgh I attended Jameson's lectures on Geology and Zoology, but they were incredibly dull. The sole effect they produced on me was the determination never as long as I lived to read a book on Geology or in any way to study the science. Yet I feel sure that I was prepared for a philosophical treatment of the subject; for an old Mr Cotton in Shropshire who knew a good deal about rocks, had pointed out to me, two or three years previously a well-known large erratic boulder in the town of Shrewsbury, called the bell-stone; he told me that there was no rock of the same kind nearer than Cumberland or Scotland, and he solemnly assured me that the world would come to an end before anyone would be able to explain how this stone came where it now lay. This produced a deep impression on me and I meditated over this wonderful stone. So that I felt the keenest delight when I first read of the action of icebergs in transporting boulders, and I gloried in the progress of Geology. Equally striking is the fact that I, though now only sixty-seven years old, heard Professor Jameson, in a field lecture at Salisbury Craigs, discoursing on a trap-dyke, with amygdaloidal margins and the strata indurated on each side, with volcanic rocks all around us, and say that it was a fissure filled with sediment from above, adding with a sneer that there were men who maintained that it had been injected from beneath in a molten condition. When I think of this lecture, I do not wonder that I determined never to attend to Geology.

From attending Jameson's lectures, I became acquainted with the curator of the museum, Mr Macgillivray,[33] who afterwards published a large and excellent book on the birds of Scotland. He had not much the appearance or manners of the gentleman. I had much interesting natural-history talk with him, and he was very kind to me. He gave me some rare shells, for I at that time collected marine mollusca, but with no great zeal.

My summer vacations during these two years were wholly given up to amusements, though I always had some book in hand, which I read with interest. During the summer of 1826, I took a long walking tour with two friends with knapsacks on our backs through North Wales. We walked thirty miles most days, including one day the ascent of Snowdon. I also went with my sister Caroline a riding tour in North Wales, a servant with saddle-bags carrying our clothes. The autumns were devoted to shooting, chiefly at Mr Owen's at Woodhouse, and at my Uncle Jos's, at Maer. My zeal was so great that I used to place my shooting boots open by my bedside when I went to bed, so as not to lose half-a-minute in putting them on in the morning; and on one occasion I reached a distant part of the Maer estate on the 20th of August for black-game shooting, before I could see: I then toiled on with the gamekeeper the whole day through thick heath and young Scotch firs. I kept an exact record of every bird which I shot throughout the whole season. One day when shooting at Woodhouse with Captain Owen,[34] the eldest son and Major Hill, his cousin, afterwards Lord Berwick, both of whom I liked very much, I thought myself shamefully used, for every time after I had fired and thought that I had killed a bird, one of the two acted as if loading his gun and cried out, 'You must not count that bird, for I fired at the same time,' and the gamekeeper perceiving the joke, backed them up. After some hours they told me the joke, but it was no joke to me for I had shot a large number of birds, but did not know how many, and could not add them to my list, which I used to do by making a knot in a piece of string tied to a button-hole. This my wicked friends had perceived.

How I did enjoy shooting, but I think that I must have been half-consciously ashamed of my zeal, for I tried to persuade myself that shooting was almost an intellectual employment; it required so much skill to judge where to find most game and to hunt the dogs well.

One of my autumnal visits to Maer in 1827 was memorable from meeting there Sir J. Mackintosh,[35] who was the best converser I ever listened to. I heard afterwards with a glow of pride that he had said, 'There is something in that young man that interests me.' This must have been chiefly due to his perceiving that I listened with much interest to everything which he said, for I was as ignorant as a pig about his subjects of history, politicks and moral philosophy. To hear of praise from an eminent person, though no doubt apt or certain to excite vanity, is, I think, good for a young man, as it helps to keep him in the right course.

My visits to Maer during these two and the three succeeding years were quite delightful, independently of the autumnal shooting. Life there was perfectly free; the country was very pleasant for walking or riding; and in the evening there was much very agreeable conversation, not so personal as it generally is in large family parties, together with music. In the summer the whole family used often to sit on the steps of the old portico, with the flower-garden in front, and with the steep wooded bank, opposite the house, reflected in the lake, with here and there a fish rising or a water-bird paddling about. Nothing has left a more vivid picture on my mind than these evenings at Maer. I was also attached to and greatly revered my Uncle Jos: he was silent and reserved so as to be a rather awful man; but he sometimes talked openly with me. He was the very type of an upright man with the clearest judgement. I do not believe that any power on earth could have made him swerve an inch from what he considered the right course. I used to apply to him in my mind, the well-known ode of Horace, now forgotten by me, in which the words 'nec vultus tyranni, &c',[36] come in.

CAMBRIDGE, 1828–1831

After having spent two sessions in Edinburgh, my father perceived or he heard from my sisters, that I did not like the thought of being a physician, so he proposed that I should become a clergyman. He was very properly vehement against my turning an idle sporting man, which then seemed my probable destination. I asked for some time to consider, as from what little I had heard and thought on the subject I had scruples about declaring my belief in all the dogmas of the Church of England; though otherwise I liked the thought of being a country clergyman. Accordingly I read with care *Pearson on the Creed*[37] and a few other books on divinity; and as I did not then in the least doubt the strict and literal truth of every word in the Bible, I soon persuaded myself that our Creed must be fully accepted. It never struck me how illogical it was to say that I believed in what I could not understand and what is in fact unintelligible. I might have said with entire truth that I had no wish to dispute any dogma; but I never was such a fool as to feel and say 'credo quia incredibile'.

Considering how fiercely I have been attacked by the orthodox it seems ludicrous that I once intended to be a clergyman. Nor was this intention and my father's wish ever formally given up, but died a natural death when on leaving Cambridge I joined the *Beagle* as Naturalist. If the phrenologists are to be trusted, I was well fitted in one respect to be a clergyman. A few years ago the Secretaries of a German psychological society asked me earnestly by letter for a photograph of myself; and some time afterwards I received the proceedings of one of the meetings in which it seemed that the shape of my head had been the subject of a public discussion, and one of the speakers declared that I had the bump of Reverence developed enough for ten Priests.

As it was decided that I should be a clergyman, it was necessary that I should go to one of the English universities and take a degree; but as I had never opened a classical book since leaving school, I found to my dismay that in the two intervening years I had actually forgotten, incredible as it may appear, almost everything which I had learnt even to some few of the Greek letters. I did not therefore proceed to Cambridge at the usual time in October, but worked with a private tutor in Shrewsbury and went to Cambridge after the Christmas vacation, early in 1828. I soon recovered my school standard of knowledge, and could translate easy Greek books, such as Homer and the Greek Testament with moderate facility.

During the three years which I spent at Cambridge my time was wasted, as far as the academical studies were concerned, as completely as at Edinburgh and at school. I attempted mathematics, and even went during the summer of 1828 with a private tutor (a very dull man) to Barmouth, but I got on very slowly. The work was repugnant to me, chiefly from my not being able to see any meaning in the early steps in algebra. This impatience was very foolish, and in after years I have deeply regretted that I did not proceed far enough at least to understand something of the great leading principles of mathematics; for men thus endowed seem to have an extra sense. But I do not believe that I should ever have succeeded beyond a very low grade. With respect to Classics I did nothing except attend a few compulsory college lectures, and the attendance was almost nominal. In my second year I had to work for a month or two to pass the Little Go, which I did easily. Again in my last year I worked with some earnestness for my final degree of B.A., and brushed up my Classics together with a little Algebra and Euclid, which latter gave me much pleasure, as it did whilst at school. In order to pass the B.A. examination, it was, also, necessary to get up Paley's[38] *Evidences of Christianity*, and his *Moral Philosophy*. This was done in a thorough manner, and I am convinced that I could have written out the whole of the *Evidences* with perfect correctness, but not of course in the clear language of Paley. The logic of this book and as I may add of his *Natural Theology* gave me as much delight as did Euclid. The careful study of these works, without attempting to learn any part by rote, was the only part of the Academical Course which, as I then felt and as I still believe, was of the least use to me in the education of my mind. I did not at that time trouble myself about Paley's premises; and taking these on trust I was charmed and convinced by the long line of argumentation. By answering well the examination questions in Paley, by doing Euclid well, and by not failing miserably in Classics, I gained a good place among the οc πολλοι, or crowd of men who do not go in for honours. Oddly enough I cannot remember how high I stood, and my memory fluctuates between the fifth, tenth, or twelfth name on the list.[39]

Public lectures on several branches were given in the University, attendance being quite voluntary; but I was so sickened with lectures at Edinburgh that I did not even attend Sedgwick's[40] eloquent and interesting lectures. Had I done so I should probably have become a geologist earlier than I did. I attended, however, Henslow's[41] lectures on Botany, and liked them much for their extreme clearness, and the admirable illustrations; but I did not study botany. Henslow used to take his pupils, including several of the older members of the University, field excursions, on foot, or in coaches to distant places, or in a barge down the river, and lectured on the rarer plants or animals which were observed. These excursions were delightful.

Although as we shall presently see there were some redeeming features in my life at Cambridge, my time was sadly wasted there and worse than wasted. From my passion for shooting and for hunting and when this failed, for riding across country I got into a sporting set, including some dissipated low-minded young men. We used often to dine together in the evening, though these dinners often included men of a higher stamp, and we sometimes drank too much, with jolly singing and playing at cards afterwards. I know that I ought to feel ashamed of days and evenings thus spent, but as some of my friends were very pleasant and we were all in the highest spirits, I cannot help looking back to these times with much pleasure.[42]

But I am glad to think that I had many other friends of a widely different nature. I was very intimate with Whitley,[43] who was afterwards Senior Wrangler, and we used continually to take long walks together. He inoculated me with a taste for pictures and good engravings, of which I bought some. I frequently went to the Fitzwilliam Gallery, and my taste must have been fairly good, for I certainly admired the best pictures, which I discussed with the old curator. I read also with much interest Sir J. Reynolds' book.[44] This taste, though not natural to me, lasted for several years and many of the pictures in the National Gallery in London gave me much pleasure; that of Sebastian del Piombo[45] exciting in me a sense of sublimity.

I also got into a musical set, I believe by means of my warm-hearted friend Herbert,[46] who took a high wrangler's degree. From associating with these men and hearing them play, I acquired a strong taste for music, and used very often to time my walks so as to hear on week days the anthem in King's College Chapel. This gave me intense pleasure, so that my backbone would sometimes shiver. I am sure that there was no affectation or mere imitation in this taste, for I used generally to go by myself to King's College, and I sometimes hired the chorister boys to sing in my rooms. Nevertheless I am so utterly destitute of an ear, that I cannot perceive a discord, or keep time and hum a tune correctly; and it is a mystery how I could possibly have derived pleasure from music.

My musical friends soon perceived my state, and sometimes amused themselves by making me pass an examination, which consisted in ascertaining how many tunes I could recognize, when they were played rather more quickly or slowly than usual. 'God save the King' when thus played was a sore puzzle. There was another man with almost as bad an ear as I had, and strange to say he played a little on the flute. Once I had the triumph of beating him in one of our musical examinations.

But no pursuit at Cambridge was followed with nearly so much eagerness or gave me so much pleasure as collecting beetles. It was the mere passion for collecting, for I did not dissect them and rarely compared their external characters with published descriptions, but got them named anyhow. I will give a proof of my zeal: one day, on tearing off some old bark, I saw two rare beetles and seized one in each hand; then I saw a third and new kind, which I could not bear to lose, so that I popped the one which I held in my right hand into my mouth. Alas it ejected some intensely acrid fluid, which burnt my tongue so that I was forced to spit the beetle out, which was lost, as well as the third one.

I was very successful in collecting and invented two new methods; I employed a labourer to scrape during the winter, moss off old trees and place [it] in a large bag, and likewise to collect the rubbish at the bottom of the barges in which reeds are brought from the fens, and thus I got some very rare species. No poet ever felt more delight at seeing his first poem published than I did at seeing in Stephen's *Illustrations of British Insects*[47] the magic words, 'captured by C. Darwin, Esq.'. I was introduced to entomology by my second cousin, W. Darwin Fox,[48] a clever and most pleasant man, who was then at Christ's College, and with whom I became extremely intimate. Afterwards I became well acquainted with and went out collecting, with Albert Way[49] of Trinity, who in after years became a well-known archaeologist; also with H. Thompson,[50] of the same College, afterwards a leading agriculturist, chairman of a great Railway, and Member of Parliament. It seems therefore that a taste for collecting beetles is some indication of future success in life!

I am surprised what an indelible impression many of the beetles which I caught at Cambridge have left on my mind. I can remember the exact appearance of certain posts, old trees and banks where I made a good capture. The pretty *Panagaeus crux-major* was a treasure in those days, and here at Down I saw a beetle running across a walk, and on picking it up instantly perceived that it differed slightly from *P. crux-major*, and it turned out to be *P. quadripunctatus*, which is only a variety or closely allied species, differing from it very slightly in outline. I had never seen in those old days Licinus alive, which to an uneducated eye hardly differs from many other black Carabidous beetles; but my sons found here a specimen and I instantly recognized that it was new to me; yet I had not looked at a British beetle for the last twenty years.

I have not as yet mentioned a circumstance which influenced my whole career more than any other. This was my friendship with Prof. Henslow. Before coming up to Cambridge, I had heard of him from my brother as a man who knew every branch of science, and I was accordingly prepared to reverence him. He kept open house once every week,[51] where all undergraduates and several older members of the University, who were attached to science, used to meet in the evening. I soon got, through Fox, an invitation, and went there regularly. Before long I became well acquainted with Henslow, and during the latter half of my time at Cambridge took long walks with him on most days; so that I was called by some of the dons 'the man who walks with Henslow'; and in the evening I was very often asked to join his family dinner. His knowledge was great in botany, entomology, chemistry, mineralogy, and geology. His strongest taste was to draw conclusions from long-continued minute observations. His judgement was excellent, and his whole mind well-balanced; but I do not suppose that anyone would say that he possessed much original genius.

He was deeply religious, and so orthodox, that he told me one day, he should be grieved if a single word of the Thirty-nine Articles were altered. His moral qualities were in every way admirable. He was free from every tinge of vanity or other petty feeling; and I never saw a man who thought so little about himself or his own concerns. His temper was imperturbably good, with the most winning and courteous manners; yet, as I have seen, he could be roused by any bad action to the warmest indignation and prompt action. I once saw in his company in the streets of Cambridge almost as horrid a scene, as could have been witnessed during the

French Revolution. Two body-snatchers had been arrested and whilst being taken to prison had been torn from the constable by a crowd of the roughest men, who dragged them by their legs along the muddy and stony road. They were covered from head to foot with mud and their faces were bleeding either from having been kicked or from the stones; they looked like corpses, but the crowd was so dense that I got only a few momentary glimpses of the wretched creatures. Never in my life have I seen such wrath painted on a man's face, as was shown by Henslow at this horrid scene. He tried repeatedly to penetrate the mob; but it was simply impossible. He then rushed away to the mayor, telling me not to follow him, to get more policemen. I forget the issue, except that the two were got into the prison before being killed.

Henslow's benevolence was unbounded, as he proved by his many excellent schemes for his poor parishioners, when in after years he held the living of Hitcham. My intimacy with such a man ought to have been and I hope was an inestimable benefit. I cannot resist mentioning a trifling incident, which showed his kind consideration. Whilst examining some pollen-grains on a damp surface I saw the tubes exserted, and instantly rushed off to communicate my surprising discovery to him. Now I do not suppose any other Professor of Botany could have helped laughing at my coming in such a hurry to make such a communication. But he agreed how interesting the phenomenon was, and explained its meaning, but made me clearly understand how well it was known; so I left him not in the least mortified, but well pleased at having discovered for myself so remarkable a fact, but determined not to be in such a hurry again to communicate my discoveries.

Dr Whewell[52] was one of the older and distinguished men who sometimes visited Henslow, and on several occasions I walked home with him at night. Next to Sir J. Mackintosh he was the best converser on grave subjects to whom I ever listened. Leonard Jenyns,[53] (grandson of the famous Soames Jenyns), who afterwards published some good essays in Natural History, often staid with Henslow, who was his brother-in-law. At first I disliked him from his somewhat grim and sarcastic expression; and it is not often that a first impression is lost; but I was completely mistaken and found him very kindhearted, pleasant and with a good stock of humour. I visited him at his parsonage on the borders of the Fens [Swaffham Bulbeck], and had many a good walk and talk with him about Natural History. I became also acquainted with several other men older than me, who did not care much about science, but were friends of Henslow. One was a Scotchman, brother of Sir Alexander Ramsay, and tutor of Jesus College; he was a delightful man, but did not live for many years.[54] Another was Mr Dawes,[55] afterwards Dean of Hereford and famous for his success in the education of the poor. These men and others of the same standing, together with Henslow, used sometimes to take distant excursions into the country, which I was allowed to join and they were most agreeable.

Looking back, I infer that there must have been something in me a little superior to the common run of youths, otherwise the above-mentioned men, so much older than me and higher in academical position, would never have allowed me to associate with them. Certainly I was not aware of any such superiority, and I remember one of my sporting friends, Turner, who saw me at work on my beetles, saying that I should some day be a Fellow of the Royal Society, and the notion seemed to me preposterous.

During my last year at Cambridge I read with care and profound interest Humboldt's *Personal Narrative*.[56] This work and Sir J. Herschel's[57] *Introduction to the Study of Natural Philosophy* stirred up in me a burning zeal to add even the most humble contribution to the noble structure of Natural Science. No one or a dozen other books influenced me nearly so much as these two. I copied out from Humboldt long passages about Teneriffe, and read them aloud on one of the above-mentioned excursions, to (I think) Henslow, Ramsay and Dawes; for on a previous occasion I had talked about the glories of Teneriffe, and some of the party declared they would endeavour to go there; but I think that they were only half in earnest. I was, however, quite in earnest, and got an introduction to a merchant in London to enquire about ships; but the scheme was of course knocked on the head by the voyage of the *Beagle*.

My summer vacations were given up to collecting beetles, to some reading and short tours. In the autumn my whole time was devoted to shooting, chiefly at Woodhouse and Maer, and sometimes with young Eyton of Eyton.[58] Upon the whole the three years which I spent at Cambridge were the most joyful in my happy life; for I was then in excellent health, and almost always in high spirits.

As I had at first come up to Cambridge at Christmas, I was forced to keep two terms after passing my final examination, at the commencement of 1831; and Henslow then persuaded me to begin the study of geology. Therefore on my return to Shropshire I examined sections and coloured a map of parts round Shrewsbury. Professor Sedgwick intended to visit N. Wales in the beginning of August to pursue his famous geological investigation amongst the older rocks, and Henslow asked him to allow me to accompany him.[59] Accordingly he came and slept at my Father's house.

A short conversation with him during this evening produced a strong impression on my mind. Whilst examining an old gravel-pit near Shrewsbury a labourer told me that he had found in it a large worn tropical Volute shell, such as may be seen on the chimney-pieces of cottages; and as he would not sell the shell I was convinced that he had really found it in the pit. I told Sedgwick of the fact, and he at once said (no doubt truly) that it must have been thrown away by someone into the pit; but then added, if really embedded there it would be the greatest misfortune to geology, as it would overthrow all that we know about the superficial deposits of the midland counties. These gravel-beds belonged in fact to the glacial period, and in after years I found in them broken arctic shells. But I was then utterly astonished at Sedgwick not being delighted at so wonderful a fact as a tropical shell being found near the surface in the middle of England. Nothing before had ever made me thoroughly realize, though I had read various scientific books, that science consists in grouping facts so that general laws or conclusions may be drawn from them.

Next morning we started for Llangollen, Conway, Bangor, and Capel Curig. This tour was of decided use in teaching me a little how to make out the geology of a country. Sedgwick often sent me on a line parallel to his, telling me to bring back specimens of the rocks and to mark the stratification on a map. I have little doubt that he did this for my good, as I was too ignorant to have aided him. On this tour I had a striking instance how easy it is to overlook phenomena, however conspicuous, before they have been observed by anyone. We spent many hours in Cwm Idwal, examining all the rocks with extreme care, as Sedgwick was anxious to find fossils in them; but neither of us saw a trace of the wonderful glacial phenomena all around us; we did not notice the plainly scored rocks, the perched boulders, the lateral and terminal moraines. Yet these phenomena are so conspicuous that, as I declared in a paper published many years afterwards in the *Philosophical Magazine*,[60] a house burnt down by fire did not tell its story more plainly than did this valley. If it had still been filled by a glacier, the phenomena would have been less distinct than they now are.

At Capel Curig I left Sedgwick and went in a straight line by compass and map across the mountains to Barmouth, never following any track unless it coincided with my course. I thus came on some strange wild places and enjoyed much this manner of travelling. I visited Barmouth to see some Cambridge friends who were reading there, and thence returned to Shrewsbury and to Maer for shooting; for at that time I should have thought myself mad to give up the first days of partridge-shooting for geology or any other science.

NOTES

It is difficult to say how much importance Darwin attached to his autobiographical fragment. The essay seems to have been undertaken as a belated recognition of a family responsibility, and may not have been prepared for publication. It is also relatively short; the selection reprinted here, approximately one-third of the text, contains valuable information not available in the more formal writings and correspondence of Darwin's friends and acquaintances. Though always readable and of great help in assessing how Darwin saw himself late in his career, it may be most notable for how much of a crowded life Darwin chose not to summarize. Begun in May 1876, when Darwin was sixty-seven, and concluded in early August, the manuscript was enlarged periodically, with some sixty-seven pages of Addenda, until the year Darwin died.

The *Autobiography* was published first as a substantial section of *Life and Letters of Charles Darwin*, edited by Francis Darwin (London: John Murray, 1887). Several blunt passages recording Darwin's attitude toward orthodox Christianity were deleted because Darwin's widow and members of his family objected to their inclusion. They were not restored until Nora Barlow, Darwin's granddaughter, published in 1958 a complete transcript of the entire manuscript, which forms part of the Darwin Collection in the University Library, Cambridge. The style throughout is clear, straightforward, and sufficient unto what needs to be said at any given moment. (I have retained many of Francis Darwin's annotations, and expanded them when necessary.)

Darwin, in the section reprinted here, offers a lively character sketch of his father, and follows it with a review of his early education, up to and including his years at Edinburgh and Cambridge Universities. Darwin did not hesitate to censure dull, incompetent or useless pedagogy, but he admitted, somewhat disarmingly, that his adolescent years were spent without a clear professional goal in mind. The voyage of *HMS Beagle* transformed his life.

1. Francis Darwin adds an ironic footnote: though his father, Charles Darwin, attended Chapel meetings with his

elder sisters, both Charles and Erasmus 'were christened and intended to belong to the Church of England; and after his early boyhood [Charles] seems usually to have gone to church and not to Mr Case's'.

2.   Revd. W. A. Leighton, who was a schoolfellow of my father's at Mr Case's school, remembers his bringing a flower to school and saying that his mother had taught him how by looking at the inside of the blossom the name of the plant could be discovered. Mr Leighton goes on, 'This greatly roused my attention and curiosity, and I inquired of him repeatedly how this could be done?' – but his lesson was naturally enough not transmissible. [FD]

William Allport Leighton (1805–99) apparently fell for more than one hoax promulgated by Charles Darwin; on one occasion, for example, he believed that Darwin could change the colours of primroses and polyanthuses by dosing. Educated at St John's College, Leighton became a Fellow at Christ's, and wrote two well-regarded books, *Flora of Shropshire* and *Lichen Flora of Great Britain*.

3.   His father wisely treated this tendency not by making crimes of the fibs, but by making light of the discoveries. [FD]

4.   Robert Darwin loaned Josiah Wedgwood II the sum of £30,000 to buy, for a second home in the country, a thousand-acre estate in the Staffordshire village of Maer, some twenty miles from Shrewsbury. The plain stone house in which this family of Wedgwood cousins lived enjoyed gardens that had been designed by Capability Brown. Josiah was Charles Darwin's uncle. Emma Wedgwood, destined to become Charles Darwin's wife, was born here as Josiah's last child (1808).

5.   Charles Darwin's portrait of his father was added in 1878 to the MS. of the *Autobiography*.

6.   The 'connection' was Henry Parker, who had married Marianne, the oldest daughter of Robert Darwin.

7.   William Petty, first Marquis of Landsdowne (1737–1805). Darwin consistently misspelled Shelburne's name as Sherburn.

8.   Erasmus Darwin, writing on 5 January 1792 to his son Robert, argued that alcohol was probably responsible for the development of both insanity and epilepsy; both diseases, he believed, were 'hereditary in some degree'.

9. This belief still survives, and was mentioned to my brother in 1884 by an old inhabitant of Shrewsbury. [FD]

10. Erasmus Darwin.

11. The title of Carlyle's book is *Reminiscences* (rather than *Remembrances*).

12. Samuel Tertius Galton (1783–1833), son of Samuel John G. Dalton.

13. *Wonders of the World* (Dublin: B. Smith, 1825).

14. Reverend Gilbert White, *Natural History and Antiquities of Selborne* (1789).

15. Samuel Parkes, *Chemical Catechism with tables, notes, illustration, and experiments* (London: Baldwar, 1822).

16. *Poco curante*: taking little care (It.).

17. He lodged at Mrs Mackay's, 11 Lothian Street. What little the records of Edinburgh University can reveal has been published in the *Edinburgh Weekly Dispatch*, 22 May 1888; and in the *St James's Gazette*, 16 February 1888. From the latter journal it appears that he and his brother Erasmus made more use of the library than was usual among the students of their time. [FD]

18. Thomas Charles Hope (1766–1844) performed 'chemical dramas' in his hugely popular chemistry courses at Edinburgh University (1799–1845). Darwin, who came into contact with him during his first year at Edinburgh, admired him greatly.

19. Andrew Duncan, Professor of Materia Medica, the man who isolated cinchonin from Peruvian bark, and a publicizer of achievements in French science, may have been unfairly characterized by Darwin on more than one occasion. Writing to Hooker, Darwin complained of one of Duncan's lectures as 'a whole cold breakfastless hour on the properties of rhubarb'.

20. Alexander Monro III (1773–1859) followed his father's profession when he decided to specialize in human anatomy. For several years he held a joint appointment in medicine and surgery with his father at the University of Edinburgh; from 1808 on he taught the whole course by himself; from 1817 on he was the sole professor. His reputation as a dull dog in the lecture-room was written about, unfavourably, by many students; *DNB* offers a withering judgement on his lack of originality. It was widely believed that he lectured from his grandfather's notes.

21. I have heard him call to mind the pride he felt at the results of the successful treatment of a whole family with tartar emetic. [FD]

22. William Francis Ainsworth (1807–96) was a physician and geologist who sometimes lectured on principles of natural classification (on one occasion Darwin was called upon for his comments), and an important member of several expeditions to Assyria.

23. Abraham Gottlob Werner (1750–1817) was a geologist who believed that all rocks of the earth's crust originated under water. This line of argument, which attracted several scientists, was known as the Neptunian theory.

24. John Coldstream (1806–63) was a physician and naturalist, and very active in both the Wernerian and Plinian Societies before moving (in part, apparently, for health reasons) from Edinburgh to Leith; he had well-developed interests in the Bible and missionary activities. He was a pioneer in the treatment of invalids and imbeciles.

25. Dr Hardie, a physician, went on several natural history trips with Darwin. He was identified by one of Darwin's teachers, James Hartley Ashworth, Professor of Zoology at Edinburgh University, as Willoughby Arding; the basis of Professor Ashworth's identification is unclear. Later, Ashworth went to Bombay on assignment; still later, Darwin noted (in a casual remark) that Arding had died young.

26. Robert Edmund Grant (1793–1874) collected specimens on the seashore with Darwin, and examined, at Darwin's request, corallines from the *Beagle*. He dedicated *Tabular View of the Primary Divisions of the Animal Kingdom* (1861) to Darwin. Huxley thought Grant's work was promising and that Grant stopped publishing prematurely.

27. The society was founded in 1823, and expired about 1848 (*Edinburgh Weekly Dispatch*, 22 May 1888). [FD]

28. Robert Jameson, mineralogist, founded both the Wernerian and Plinian Societies (1808 and 1823 respectively). He held the post of Keeper of the Museum at Edinburgh (1804–54).

29. James Phillips Kay-Shuttleworth (1804–77), physician, was the first Secretary of the Committee of the Council of

Education, and – more than any other individual – helped to shape what was to become England's system of public education. A noted philanthropist and champion of poor people in their efforts to improve health conditions, Kay-Shuttleworth did yeoman service in advancing the cause of scientific research that might be put to practical uses.

30. John James Audubon (1780–1851), ornithologist, spent the year 1826–27 in England. His aim, which he achieved, was to obtain enough subscribers to enable him to begin publication of 435 coloured plates. These he called *The Birds of America* (published in 1838).

31. Charles Waterton (1782–1865) was a naturalist whose *Wanderings in South America* (1825) won him international recognition, and who had the pleasure of welcoming Darwin to Walton Hall, near Pontefract, Yorkshire. He was, in fact, a genial host, quarrelling only on rare occasions with those who questioned his data or the conclusions he drew from them (e.g. Audubon, who disagreed with his views on how vultures found their food).

32. Leonard Horner (1785–1864), Scottish geologist and educational reformer, became a partner in his father's business (linen draper); helped to found the Edinburgh School of Arts for the instruction of mechanics; and served as a Factory Commissioner (1833–60), taking a special interest in the employment of children in factories.

33. William Macgillivray (1796–1852) held several distinguished appointments, among them that of Conservator of the Royal College of Surgeons Museum, Edinburgh (1831–41) and that of Professor of Natural History, Aberdeen (through most of the 1840s). His major book, *A History of British Birds* (1837–52), won praise from Audubon (with whom he collaborated on other works) for being the finest such study published during the century.

34. Mr Owen (later promoted to Major) was invited to Darwin's funeral as a 'personal friend'.

35. Sir James Mackintosh (1765–1832) studied medicine at the University of Edinburgh, but went on to achieve fame as a philosopher, statesman and orator in both England and India. His second marriage, to Catherine Allen, made him a relative

of Charles Darwin, since Josiah Wedgwood had married another Allen sister. A chicken bone in his throat was the unfortunate cause of his death.

36.   Justum et tenacem propositi virum
Non civium ardor prava jubentium,
Non vultus instantis tyranni
Mente quatit solida. [FD]

This quotation, from Horace's *Odes*, III, 1–4, is translated thus by Gavin de Beer: 'Just and steady-purposed man, whose mind is proof against clamouring mobs, despot's grimaces, the South wind of the Adriatic, or the crack and fall of Jupiter's heavens.' *Charles Darwin / Thomas Henry Huxley / Autobiographies* (Oxford: Oxford University Press, 1983), p. 114.

37.   Most likely a reference to William Hodge Mill (1792–1853), *Analysis of the Exposition of the Creed by Pearson*. Bishop John Pearse lived from 1613 to 1686.

38.   William Paley (1743–1805) published in 1785 his lectures at Christ's College, Cambridge, under the title *The Principles of Moral and Political Philosophy*, and in 1794 his *View of the Evidences of Christianity*, destined to become his most famous and controversial book. (Its line of argument seriously interfered with the advancement of his career within the Anglican Church.)

39.   Tenth in the list of January 1831. [FD]

40.   Adam Sedgwick: see below, pp. 57–8.

41.   John Stevens Henslow: see below, pp. 169–70.

42.   I gather from some of my father's contemporaries that he has exaggerated the Bacchanalian nature of these parties. [FD]

43.   The Revd C. Whitley, Canon of Durham, formerly Reader in Natural Philosophy in Durham University. [FD] Charles Thomas Whitley (1808–95), who reached the top of the Honours Examination List at Cambridge, shared an interest in fine art with Darwin. (Darwin's interest in painting was sincere and probably more extensive and discriminating than his diffidence would allow him to admit.)

44.   Sir Joshua Reynolds (1723–92) collected the addresses that he gave annually at the Royal Academy after its founding in 1768. (He served as its first President.) Published under the

title *Discourses*, the lectures, because of their good sense and elegant style, rapidly became classics of art criticism.

45. Sebastian del Piombo (1485–1517) was a member of Raphael's circle of artists (1510–15), but the more important phase of his career came when he collaborated with Michelangelo. (The two men admired each other's works.) Entrusted at the Vatican with the role of keeper of the Papal Seal ('piombino' means 'lead seal'), Sebastian produced relatively few works in later life.

46. The late John Maurice Herbert, County Court Judge of Cardiff and the Monmouth Circuit. [FD]

47. James Francis Stephens (1792–1852) was a vigorous collector of data in the fields of electricity, animals and entomology. His *Illustrations of British Entomology* (twelve volumes, 1827–46) was published under the sponsorship of the Trustees of the British Museum, and went through several editions.

48. William Darwin Fox (1805–80), Darwin's second cousin, began the study of entomology with Darwin. A lengthy correspondence between the two is of some importance in the lives of both men.

49. Albert Way (1805–74), an antiquarian, collected beetles with Darwin, and provided Darwin with much-appreciated information on the breeds of various horses.

50. Afterwards Sir H. Thompson, first baronet. [FD].

51. Henslow's Friday evenings were discontinued in 1836.

52. William Whewell (1794–1866), astronomer and philosopher, was the Master of Trinity College, Cambridge. His opposition to the teachings of Darwin led him to refuse to allow a copy of *The Origin of Species* to become part of the library of Trinity.

53. Leonard Jenyns (1800–93), later Blomefield (he changed his name on inheritance), was Henslow's brother-in-law, and later became the Vicar of Swaffham Bulbeck, Cambridgeshire. Darwin, after a negative first impression, came to like him, and asked him to describe the fish specimens gathered during the voyage of the *Beagle*. (Jenyns had almost accepted the offer of the post of naturalist on the *Beagle* before it was offered to Darwin.) Jenyns' *Chapters in My Life* (1887) remains a vivid memoir to this day. (Contrary to Francis Darwin's belief, he was not the grandson of the better-known Soames Jenyns.)

54. Marmaduke Ramsay (? –1831), a tutor at Jesus, was planning to accompany Darwin on a projected trip to Tenerife when he unexpectedly died. Darwin was severely shocked when he received the news.

55. Richard Dawes (1793–1867), a tutor at Emmanuel College, would have joined Darwin and Ramsay if the trip to the Canaries had not been cancelled as a consequence of Ramsay's death.

56. Baron Friedrich Heinrich Alexander von Humboldt (1769–1859), famous German naturalist and traveller, published *Personal Narrative of Travels to the Equinoctial Regions of the New Continent During the Years 1799–1804* (7 volumes, 1814–29, various imprints).

57. Sir John Frederick William Herschel (1792–1871), perhaps the best-known British astronomer of the nineteenth century, published *Preliminary Discourse on the Study of Natural Philosophy* in 1830. Intended to serve as the opening volume of Lardner's *Cabinet Cyclopaedia*, it achieved instant fame as a model of style. It was subsequently translated into several languages.

58. Thomas Campbell Eyton (1809–80), ornithologist, specialized in skeletal variations. In 1839 he examined the birds brought back from the *Beagle* voyage, and his findings were printed as an appendix to Part III of *Zoology of Beagle*. He frequently hunted and fished with Darwin.

59. A telling anecdote about this tour of Wales is recorded by Darwin in a letter to Professor Hughes (see below, pp. 55–7).

60. Charles Darwin, 'Notes on the Effects produced by the Ancient Glaciers of Caernarvonshire, and on the Boulders transported by Floating Ice', *Philosophical Magazine*, vol. XXI (1842).

'The Walking Tour in North Wales', in *The Life and Letters of the Reverend Adam Sedgwick [I]*, edited by John Willis Clark and Thomas McKenny Hughes (Cambridge: Cambridge University Press, 1890), vol. I, pp. 379–81 [1831]

For two or three weeks, at the commencement of this tour [August 1831], Sedgwick was accompanied by Charles Darwin, then a young man of twenty-two. It is provoking that neither should have written down his impressions of the other at the time; for it is evident that from this time forward Sedgwick took a keen interest in him. In 1835 [1 November], while Darwin was absent on board *The Beagle*, Sedgwick wrote to Dr Butler of Shrewsbury:[1] 'His [Dr Darwin's] son is doing admirable work in South America, and has already sent home a collection above all price. It was the best thing in the world for him that he went out on the voyage of discovery. There was some risk of his turning out an idle man, but his character will be now fixed, and if God spares his life he will have a great name among the Naturalists of Europe.' In after life, though they differed widely, Sedgwick always spoke of his geological pupil, as he may be termed, with cordiality and kindness; and Darwin, replying to a note received from Sedgwick not very long before his death, could write: 'I am pleased that you remember my attending you in your excursions in 1831. To me, it was a memorable event in my life: I felt it a great honour, and it stimulated me to work, and made me appreciate the noble science of geology.' In 1875, in answer to an inquiry from Professor Hughes, Darwin wrote down all he could remember about the tour of 1831.

DOWN, BECKENHAM, KENT,
May 24, 1875.

My dear Sir,

I understand from my son that you wish to hear about my short geological tour with Professor Sedgwick in North Wales during the summer of 1831; but it is so long ago that I can tell you very little.

As I desired to learn something about Geology, Professor Henslow asked Sedgwick to allow me to accompany him on his tour, and he assented to this in the readiest and kindest manner. He came to my father's house at Shrewsbury, and I remember how spirited and amusing his conversation was during the whole evening; but he talked so much about his health and uncomfortable feelings that my father, who was a doctor, thought that he was a confirmed hypochondriac.

We started next morning, and after a day or two he sent me across the country in a line parallel to his course, telling me to collect specimens of the rocks, and to note the stratification. In the evening he discussed what I had seen; and this of course encouraged me greatly, and made me exceedingly proud; but I now suspect that it was done merely for the sake of teaching me, and not for anything of value which I could have told him. I remember one little incident. We left Conway early in the morning, and for the first two or three miles of our walk he was gloomy, and hardly spoke a word. He then suddenly burst forth: 'I know that the d——d fellow never gave her the sixpence. I'll go back at once;' and turned round to return to Conway. I was amazed, for I never heard before, or since, anything like an oath from him. On inquiry I found that he was convinced that the waiter had not given to the chambermaid the sixpence which he had left for her. He had no reason whatever, excepting that he thought the waiter 'an ill-looking fellow'. On my hinting that he could hardly accuse a man of theft on such grounds, he consented to proceed, but for some time he grumbled and growled. At last his brow cleared, and we had a delightful day, and he was as energetic as on all former occasions in climbing the mountains. We spent nearly a whole day in Cwm Idwal examining the rocks carefully, as he was very desirous to find fossils.

I have often thought of this day as a good instance of how easy it is for any one to overlook new phenomena, however conspicuous they may be. The valley is glaciated in the plainest manner, the rocks being mammillated, deeply scored, with many perched boulders, and well-defined moraines; yet none of these phenomena were observed by Professor Sedgwick, nor of course by me. Nevertheless they are so plain, that, as I saw in 1842, the presence of a glacier filling the valley would have rendered the evidence less distinct.

Shortly afterwards I left Professor Sedgwick, and struck across the country in another direction, and reported by letter what I saw. In his answer he discussed my ignorant remarks in his usual generous and frank manner. I am sorry to say that I can tell you nothing more about our little tour.

I find that I have kept only one letter from Professor Sedgwick, which he wrote after receiving a copy of my *Origin of Species*. His judgement naturally does not seem to me quite a fair one, but I think that the letter is characteristic of the man, and you are at liberty to publish it if you should so desire.

Believe me, my dear Sir,
Yours sincerely,
CHARLES DARWIN.

NOTES

Adam Sedgwick (1785–1873) became one of the founders of the science of geology, despite the fact that he had been appointed the Woodwardian Professor at Cambridge University while

still ignorant of much of the content of the science. A popular
lecturer, he earned an international reputation as a consequence of his determined efforts to conduct original research
and gather disciples. All who wrote about him considered him
to be an engaging human being.

His notorious response to Darwin's gift of a copy of *The Origin of Species*, reprinted here, signalled a breakdown in an old
friendship and the entrenchment of an old fogeyism that could
not help but distress Darwin. Sedgwick's contributions to his
field, mostly in the form of more than fifty papers rather than a
complete book, were significant, and well respected; but Sedgwick's anger at what he thought deluded thinking on Darwin's
part must also be considered along with his slowness, even as
early as the 1840s, to accept (or even to consider objectively)
the validity of conclusions reached on the basis of research
conducted by other scientists, e.g. Sir Charles Lyell, Louis
Agassiz and William Buckland.

Sedgwick's unhappiness at the sight of a disciple gone
wrong must be balanced against Darwin's appreciation of his
role as mentor. Darwin's pleasant memories of a geological
excursion to North Wales spent in the company of Sedgwick,
who flattered him by treating him as a fellow scientist whose
views were worth inquiring into (August 1831), helped to com-
pensate for the time that he believed had been wasted by
incompetent or uninteresting teachers elsewhere. (On his
return from Wales Darwin was to discover a communication
from Henslow that encouraged him to apply for the position of
naturalist aboard *HMS Beagle*.)

1.   The Rev. Samuel Butler I (1774–1839) was not only the
Headmaster of Shrewsbury School (1798–1836) at the time Darwin was in attendance there, but served as Bishop of Lichfield
and Coventry in his final three years.

'Joining Captain Fitz-Roy on *HMS Beagle'*, *The Correspondence of Charles Darwin,* vol. 1: *1821–1836* (Cambridge: Cambridge University Press, 1989), pp. 127–9, 131–5, 140–4, 146–7, 154–5, 162–4, 178 [1831]

From George Peacock[1] to J. S. Henslow [6 or 13 August 1831]

My dear Henslow

Captain Fitz Roy is going out to survey the southern coast of Terra del Fuego, & afterwards to visit many of the South Sea Islands & to return by the Indian Archipelago: the vessel is fitted out expressly for scientific purposes, combined with the survey; it will furnish therefore a rare opportunity for a naturalist & it would be a great misfortune that it should be lost.

An offer has been made to me to recommend a proper person to go out as a naturalist with this expedition; he will be treated with every consideration; the Captain is a young man of very pleasing manners (a nephew of the Duke of Grafton), of great zeal in his profession & who is very highly spoken of; if Leonard Jenyns[2] could go, what treasures he might bring home with him, as the ship would be placed at his disposal, whenever his enquiries made it necessary or desirable; in the absence of so accomplished a naturalist, is there any person whom you could strongly recommend: he must be such a person as would do credit to our recommendation.

Do think on this subject: it would be a serious loss to the cause of natural science, if this fine opportunity was lost.

The ship sails about the end of Sept.

Poor Ramsay![3] what a loss to us all & particularly to you.

Write immediately & tell me what can be done.

Believe me | My dear Henslow | Most truly yours | George Peacock

7. Suffolk Street | Pall Mall East

My dear Henslow

I wrote this letter on Saturday, but I was too late for the Post: What a glorious opportunity this would be for forming collections for our museums: do write to me immediately & take care that the opportunity is not lost.

Believe me | My dear Henslow | Most truly yours | Geo Peacock

7. Suffolk St | Monday

From J. S. Henslow   24 August 1831

Cambridge
24 Aug 1831

My dear Darwin,

Before I enter upon the immediate business of this letter, let us condole together upon the loss of our inestimable friend poor Ramsay, of whose death you have undoubtedly heard long before this. I will not now dwell upon this painful subject as I shall hope to see you shortly fully expecting that you will eagerly catch at the offer which is likely to be made you of a trip to Terra del Fuego & home by the East Indies – I have been asked by Peacock who will read & forward this to you from London to recommend him a naturalist as companion to Capt Fitzroy employed by Government to survey the S. extremity of America – I have stated that I consider you to be the best qualified person I know of who is likely to undertake such a situation – I state this not on the supposition of yr. being a *finished* Naturalist, but as amply qualified for collecting, observing, & noting any thing worthy to be noted in Natural History. Peacock has the appointment at his disposal & if he can not find a man willing to take the office, the opportunity will probably be lost – Capt. F. wants a man (I understand) more as a companion than a mere collector & would not take any one however good a Naturalist who was not recommended to him likewise as a *gentleman*. Particulars of salary &c I know nothing. The Voyage is to last 2 yrs. & if you take plenty of Books with you, any thing you please may be done. – You will have ample opportunities at command. – In short I suppose there never was a finer chance for a man of zeal & spirit. Capt F. is a young man. What I wish you to do is instantly to come to Town & consult with Peacock (at No. 7 Suffolk Street, Pall Mall East, or else at the University Club) & learn further particulars. Don't put on any modest doubts or fears about your disqualifications for I assure you I think you are the very man they are in search of – so conceive yourself to be tapped on the Shoulder by your Bum-Bailiff[4] & affecte friend | J. S. Henslow

The expedn. is to sail on 25 Sept: (at earliest) so there is no time to be lost.

To J. S. Henslow   30 [August 1831]

Shrewsbury
Tuesday 30th.

My dear Sir

Mr. Peacock's letter arrived on Saturday, & I received it late yesterday evening. – As far as my own mind is concerned, I should I think, *certainly* most gladly have accepted the opportunity, which you so kindly have offered me. – But my Father, although he does not decidedly refuse me, gives such strong advice against going – that I should not be comfortable, if I did not follow it. – My Father's objections are these; the unfitting me to settle down as a clergyman. – My little habit of seafaring. – The *shortness of the time* & the chance of my not suiting Captain Fitzroy. – It is certainly a very serious objection, the very short time for all my preparations, as not only body but mind wants making up for such an undertaking. – But if it had not been for my Father, I would have taken all risks.

What was the reason, that a Naturalist was not long ago fixed upon? – I am very much obliged for the trouble you have had about it – there certainly could not have been a better opportunity. – I shall come up in October to Cambridge, when I long to have some talk with you. – I will write to Mr. Peacock at Denton, (in Durham?) but his direction is written so badly, that even with the assistance of the Post office, I am not certain about it. – Would you therefore be so kind, if you know his or C. Fitzroy's direction, would you send one line to the same effect. – My trip with Sedgwick answered most perfectly. – I did not hear of poor Mr. Ramsay's loss till a few days before your letter. I have been lucky hitherto, in never losing any person for whom I had any esteem or affection. My acquaintance, although very short, was sufficient to give me those feelings in a great degree. – I can hardly make myself believe he is no more. – He was the finest character I ever knew.

Yours most sincerely | my dear Sir. Chas. Darwin.

I have written to Mr. Peacock, & I mentioned that I have asked you to send one line in the chance of his not getting my letter. – I have also asked him to communicate with Cap. Fitzroy. – Even if I was to go, my Father disliking would take away all energy, & I should want a good stock of that. – Again I must thank you; it adds a little to the heavy, but pleasant load of gratitude which I owe to you.

From R. W. Darwin to Josiah Wedgwood II[5] 30–1 August 1831

Salop
30 August 31.

Dear Wedgwood
    ...Charles will tell you of the offer he has had made to him of going for a voyage of discovery for 2 years. – I strongly object to it ⟨on var⟩ious grounds, but I will not detail my reasons that he may have your unbiased opinion on the subject, & if you think differently from me I shall wish him to follow your advice.
    Dear Wedgwood yours affectionly | R W Darwin....

Wednesday 31.

    Charles has quite given up the idea of the voyage.

To R. W. Darwin   31 August [1831]

[Maer]
August 31

My dear Father
    I am afraid I am going to make you again very uncomfortable. – But upon consideration, I think you will excuse me once again stating my opinions on the offer of the Voyage. – My excuse & reason is, is the different way all the Wedgwoods view the subject from what you & my sisters do.
    I have given Uncle Jos. what I fervently trust is an accurate & full list of your objections, & he is kind enough to give his opinion on all. – The list & his answers will be enclosed.[6] – But may I beg of you one favour. it will be doing me the greatest kindness, if you will send me a decided answer, yes or no. – If the latter, I should be most ungrateful if I did not implicitly yield to your better judgement & to the kindest indulgence which you have shown me all through my life. – & you may rely upon it I will never mention the subject again. – if your answer should be yes; I will go directly to

Henslow & consult deliberately with him & then come to Shrewsbury. – The danger appears to me & all the Wedgwoods not great. – The expense can not be serious, & the time I do not think anyhow would be more thrown away, than if I stayed at home. – But pray do not consider, that I am so bent on going, that I would for one *single moment* hesitate, if you thought, that after a short period, you should continue uncomfortable.

I must again state I cannot think it would unfit me hereafter for a steady life. – I do hope this letter will not give you much uneasiness. – I send it by the Car tomorrow morning, if you make up your mind directly will you send me an answer on the following day, by the same means. – If this letter should not find you at home, I hope you will answer as soon as you conveniently can.

I do not know what to say about Uncle Jos.' kindness, I never can forget how he interests himself about me.

Believe me my dear Father | Your affectionate son | Charles Darwin.

PS. Frank would be much obliged if you would forward the Crockery to the Hill.

(1)   Disreputable to my character as a Clergyman hereafter
(2)   A wild scheme
(3)   That they must have offered to many others before me, the place of Naturalist
(4)   And from its not being accepted there must be some serious objection to the vessel or expedition
(5)   That I should never settle down to a steady life hereafter
(6)   That my accommodations would be most uncomfortable
(7)   That you should consider it as again changing my profession
(8)   That it would be a useless undertaking

From Josiah Wedgwood II to R. W. Darwin   31 August 1831

Maer
31 August 1831

My dear Doctor

I feel the responsibility of your application to me on the offer that has been made to Charles as being weighty, but as you have desired Charles to consult me I cannot refuse to give the result of such consideration as I have been able to give it. Charles has put down what he conceives to be your principal objections & I think the best course I can take will be to state what occurs to me upon each of them.

1 – I should not think that it would be in any degree disreputable to his character as a clergyman. I should on the contrary think the offer honourable to him, and the pursuit of Natural History, though certainly not professional, is very suitable to a Clergyman.

2 – I hardly know how to meet this objection, but he would have definite objects upon which to employ himself and might acquire and strengthen, habits of application, and I should think would be as likely to do so in any way in which he is likely to pass the next two years at home.

3 – The notion did not occur to me in reading the letters & on reading them again with that object in my mind I see no ground for it.

4 – I cannot conceive that the Admiralty would send out a bad vessel on such a service. As to objections to the expedition, they will differ in each man's case & nothing would, I think, be inferred in Charles's case if it were known that others had objected.

5 – You are a much better judge of Charles's character than I can be. If, on comparing this mode of spending the next two years, with the way in which he will probably spend them if he does not accept this offer, you think him more likely to be rendered unsteady & unable to settle, it is undoubtedly a weighty objection – Is it not the case that sailors are prone to settle in domestic and quiet habits.

6 – I can form no opinion on this further than that, if appointed by the Admiralty, he will have a claim to be as well accommodated as the vessel will allow.

7 – If I saw Charles now absorbed in professional studies I should probably think it would not be advisable to interrupt them, but this is not, and I think will not be, the case with him. His present pursuit of knowledge is in the same track as he would have to follow in the expedition.

8 – The undertaking would be useless as regards his profession, but looking upon him as a man of enlarged curiosity; it affords him such an opportunity of seeing men and things as happens to few.

You will bear in mind that I have had very little time for consideration & that you & Charles are the persons who must decide.

I am | My dear Doctor | Affectionately yours | Josiah Wedgwood

From R. W. Darwin to Josiah Wedgwood II   1 September 1831

[Shrewsbury]
1 Sept 1831

Dear Wedgwood,

Charles is very grateful for your taking so much trouble & interest in his plans. I made up my mind to give up all objections, if you should not see it in the same view as I did.

Charles has stated my objections quite fairly & fully – if he still continues in the same mind after further enquiry, I will give him all the assistance in my power.

Many thanks for your kindness – yours | affectionly | R W Darwin

To Francis Beaufort[7]   1 September [1831]

Shrewsbury
September the 1st.

Sir

I take the liberty of writing to you according to Mr. Peacock's desire to acquaint you with my acceptance of the offer of going with Capt Fitzroy. Perhaps you may have received a letter from Mr. Peacock, stating my refusal; this was owing to my Father not at first approving of the plan, since which time he has reconsidered the subject: & has given his consent & therefore if the appointment is not already filled up – I shall be very happy to have the honour of accepting it. – There has been some delay owing to my being in Wales, when the letter arrived. – I set out for Cambridge tomorrow morning, to see Professor Henslow: & from thence will proceed immediately to London....

To Susan Darwin   [5 September 1831]

17 Spring Gardens London
Monday

I have so little time to spare that I have none to waste in rewriting letters so that you must excuse my bringing up the other with me & altering it. – The last letter was written in the morning. In middle of day Wood[8] received a letter from C. Fitzroy, which I must say was *most* straightforward & **gentlemanlike**, but so much against my going, that I immediately gave up the scheme. – & Henslow did the same: saying that he thought Peacock has acted *very wrong* in misrepresenting things so much. – I scarcely thought of going to Town, but here I am & now for more details & much more promising ones. – Cap Fitzroy is in town & I have seen him; it is no use attempting to praise him as much as I feel inclined to do, for you would not believe me. – One thing I am certain of, nothing could be more open & kind than he was to me. – It seems he had promised to take a friend[9] with him, who is in office & cannot go. – & he only

received the letter 5 minutes before I came in: & this makes
things much better for me, as want of room of was one of
F's greatest objections. – He offers me to go share in every
thing in his cabin, if I like to come; & every sort of accommodation than I can have but they will not be numerous. –
He says that nothing would be so miserable for him as having me with him if I was uncomfortable, as in small vessel
we must be thrown together, & thought it his duty to state
every thing in the worst point of view: I think I shall go on
Sunday to Plymouth to see the Vessel. – There is something
most extremely attractive in his manners, & way of coming
straight to the point. – If I live with him he say I must live
poorly, no wine & the plainest dinners. – The scheme is not
certainly so good as Peacock describes: C. F. advises me not
make my mind quite yet: but that seriously, he thinks it will
have much more pleasure than pain for me.

The Vessel does not sail till the 10th of October. – It contains 60 men 5 or 6 officers &c. – but is a small vessel – it will
probably be out nearly 3 years. – I shall pay to mess the
same as Captain does himself 30£ per annum, & Fitzroy
says if I spend including my outfitting 500 it will be beyond
the extreme. – But now for still worse news, the round the
world is not *certain*, but the chance, most excellent: till that
point is decided I will not be so. – And you may believe after
the many changes I have made, that nothing but my reason
shall decide me.

Fitzroy says the stormy sea is exaggerated, that if I do not
chuse to remain with them, I can at any time get home to
England, so many vessels sail that way & that during bad
weather (probably 2 months) if I like, I shall be left in some
healthy, safe & nice country: that I shall alway have assistance – that he has many books, all instrument, guns, at my
service – that the fewer & cheaper clothes I take the better.

The manner of proceeding will just suit me. They anchor
the ship & then remain for a fortnight at a place.

I have made Cap Beaufort perfectly understand me: he
says if I start & do not go round the world, I shall have good
reason to think myself deceived. – I am to call the day after

tomorrow, & if possible to receive more certain instructions. – The want of room is decidedly the most serious objection: but Cap Fitz. (probably owing to Wood's letter) seems determined to make me comfortable as he possibly can. – I like his manner of proceeding. – He asked me at once. – 'shall you bear being told that I want the cabin to myself? when I want to be alone. – If we treat each other this way, I hope we shall suit, if not probably we should wish each other at the Devil' We stop a week at the Madeira islands: & shall see most of big cities in S. America. C. Beaufort is drawing up the track through the South Sea.

I am writing in great hurry: I do not know whether you take interest enough to excuse treble postage. – I hope I am judging reasonably, & not through prejudice about Cap. Fitz: if so I am sure we shall suit. – I dine with him today. – I could write great deal more if I thought you liked it, & I had at present time. – There is indeed a tide in the affairs of men,[10] & I have experienced it, & I had *entirely* given it up till 1 today: Love to my Father, dearest Susan | good bye, Chas. Darwin

To J. S. Henslow   [5 September 1831]

[17 Spring Gardens] London
Monday

My dear Sir,

Gloria in excelsis is the most moderate beginning I can think of. – Things are more prosperous than I should have thought possible. – Cap. Fitzroy is every thing that is delightful, if I was to praise half so much as I feel inclined, you would say it was absurd, only once seeing him. – I think he really wishes to have me. – He offers me to mess with him & he will take care I have such room as is possible. – But about the cases he says I must limit myself: but then he thinks like a sailor about size: Cap. Beaufort says I shall be upon the boards & then it will only cost me like other officers. – Ship sails 10th of October: spends a week at Madeira islands: & then Rio de Janeiro. – They all think most extremely probable, home by the Indian Archipelago: but till that is decided, I will not be so.

What has induced Cap. Fitzroy to take a better view of the case is; that Mr. Chester, who was going as a friend, cannot go: so that I shall have his place in every respect. – Cap Fitzroy has good stock of books, many of which were in my list, & rifles &c so that the outfit will be much less expensive than I supposed. – The vessel will be out 3 years. I do not object, so that my Father does not. – On Wednesday I have another interview with Cap. Beaufort, & on Sunday most likely go with Cap. Fitzroy to Plymouth. – So I hope you will keep on thinking on the subject, & just keep memoranda of what may strike you. – I will call most probably on Mr Burchill & introduce myself. – I am in Lodgings at 17, Spring Gardens.

You cannot imagine anything more pleasant, kind & open than Cap. Fitzroy's manners were to me. – I am sure it will be my fault, if we do not suit.

What changes I have had: till one today I was building castles in the air about hunting Foxes in Shropshire, now Lamas in S America. – There is indeed a tide in the affairs of men. – If you see Mr Wood, remember me most kindly to him.

Good bye, my dear Henslow | Yours most sincere friend | Chas Darwin

Excuse this letter in such a hurry.

To Susan Darwin   [6 September 1831]

17, Spring Gardens
Tuesday

My dear Susan

...I write all this as if it was settled but it is not more than it was. – excepting that from Cap. FitzRoy wishing me so much to go, & from his kindness I feel a predestination I shall start. – I spent a very pleasant evening with him yesterday: he must be more than 23 old. He is of a slight figure, & a dark but handsome edition of Mr. Kynaston.[11] – & according to my notions preeminently good manners: He is all for Economy excepting on one point, viz fire arms, he recommends me strongly to get a case of pistols like his which cost 60 £!!, & never to go on shore anywhere without loaded ones – & he is doubting about a rifle. – He says I cannot appreciate the luxury of fresh meat here. – Of course I shall buy nothing till every thing is settled: but I work all day long at my lists, putting in & striking out articles. – This is the first really cheerful day I have spent since I received the letter, & it all is owing to the sort of involuntary confidence I place in my beau ideal of a Captain.

We stop at Teneriffe. His object is to stop at as many places as possible. He takes out 20 Chronometers[12] & it will be a 'sin' not to settle the longitudes. He tells me to get it down on writing at ye Admiralty that I have the free choice to leave, as soon & wherever I like: I daresay you expect I shall turn back at the Madeira: if I have a morsel of stomach left, I won't give up. – Excuse my so often troubling & writing, the one is of great utility, the other a great amusement to me. – Most likely I shall write tomorrow.

Love to my Father. – Dearest Susan | C. Darwin

Answer by return of post.

As my instruments want altering send my things by the Oxonian, ye same night.

To Susan Darwin   [9 September 1831]

[17 Spring Gardens]
Friday Morning

My dear Susan

...To explain things from the very beginning; Cap Fitz first wished to have naturalist & then he seems to have taken a sudden horror of the chances of having somebody he should not like on board the Vessel: he confesses, his letter to Cambridge was to throw cold water on the scheme. – I don't think we shall quarrel about politics although Wood (as might be expected from a Londonderry) solemnly warned Fitzroy that I was a whig. – Cap Fitz was before Uncle Jos. – he said 'now your friends will tell you a sea Captain is the greatest brute on the face of the creation; I do not know how to help you in this case, except by hoping you will give me a trial.' – How one does change. – I actually now wish the voyage was longer before we touched Land. I feel my blood run cold at the quantity I have [to] do. – Every body seems ready to assist me. The Zoological want to make me a corresponding member; all this I can construe without crossing the Equator: – But one friend is quite invaluable, viz a Mr Yarrell,[13] a stationer & excellent naturalist: he goes to the shops with me & bullies about prices (not that I yet buy). Hang me if I give 60 £ for pistols. . . .

To Susan Darwin   [14 September 1831]

Devonport.
Wednesday Evening

My dear Susan

I arrived here yesterday evening: after a very prosperous sail of three days from London. – I suppose breathing the same air as a sea Captain is a sort of a preventive: for I scarcely ever spent three pleasanter days. – Of course there were a few moments of giddiness, as for sickness I utterly scorn the very name of it. – There were 5 or 6 very agreeable people on board, & we formed a table & stuck together, & most jolly dinners they were. – Cap. Fitz. took a little Midshipman (who by the way knows Sir F. Darwin, his name is Musters[14]) & you cannot imagine anything more kind & good humoured than the Captain's manners were to him. – Perhaps you thought I admired my beau ideal of a Captain in my former letters: all that is quite a joke to what I now feel. – Every body praises him (whether or no they know my connection with him), & indeed, judging from the little I have seen of him, he well deserves it. – Not that I suppose it is likely that such violent admiration – as I feel for him –

can possibly last. – No man is a hero to his valet, as the old saying goes. – & I certainly shall be in much the same predicament as one.

The vessel is a very small one; three masted; & carrying 10 guns: but every body says it is the best sort for our work, & of its class it is an excellent vessel: new, but well tried, & $\frac{1}{2}$ again the usual strength. – The want of room is very bad, but we must make the best of it. – I like the officers (as Cap. F. says, they would not do for St. James, but they are evidently very intelligent, active determined set of young fellows. – I keep on balancing accounts; there are several contra's which I did not expect, but on the other hand the pro's far outweigh them.

The time of sailing keeps on receding in a greater ratio, than the present time draws on: I do not believe we shall sail till the 20th of October. – I am exceedingly glad of this, as the number of things I have got to do is quite frightful. – I do not think I can stay in Shrewsbury more than 4 days. – I leave Plymouth on Friday and shall be in Cam: at the end of next week.

I found the money at the Bank, & am much obliged to my Father for it. – My spirits about the voyage are like the tide, which runs one way & that is in favour of it, but it does so by a number of little waves, which may represent all the doubts & hopes that are continually changing in my mind. After such a wonderful high wrought simile I will write no more. So good bye, my dear Susan | Yours C. Darwin

Love to my Father.

To W. D. Fox   19 [September 1831]

17 Spring Gardens (& here I shall remain till I start)
Monday 19th
My dear Fox

I returned from my expedition to see the Beagle at Plymouth on Saturday & found your most welcome letter on my table. – It is quite ridiculous what a very long period these last 20 days have appeared to me, certainly much more than as many weeks on ordinary occasions. – this will account for my not recollecting how much I told you of my plans, therefore I will begin a novo. – The expedition, under the command of Cap FitzRoy is fitted out principally for completing a survey of the S. parts of S America. The western shores of these parts have been well done by Cap King,[15] under whom Fitzroy went out second in command. We accordingly shall principally work on the Eastern coast of Patagonia from Rio de Plata to Sts of Magellan. – The second object is to ascertain the longitudes of several places, more accurately than they are at present, & to carry a series of them round the world. – The expedition is entirely a government affair.

My appointment is not a very regular affair, as the only thing the Admiralty have done is putting me on the books for Victuals, value 40£ per annum. – I have some thoughts of having it taken off again. I should certainly do so, if I thought it would give me a more absolute disposal of my collection, when I return to England. – But on the whole it is a grand & fortunate opportunity; there will be so many things to interest me – fine scenery & an endless occupation & amusement in the different branches of Nat: History: then again navigation & meteorology will amuse me on the voyage, joined to the grand requisite of there being a pleasant set of officers, & as far as I can judge this is certain. – On the other hand there is very considerable risk to one's life & health, & the leaving for so very long a time so many people whom I dearly love is oftentimes a feeling so painful, that it requires all my resolution to overcome it. – But every thing is now settled & before the 20th of Octr I trust to be on the broad sea. – My objection to the vessel is its smallness, which cramps one so for room for packing my own body & all my cases &c &c. – As to its safety I hope the Admiralty are the best judges; to a landsman's eye she looks very small. – She is a 10 gun, 3 masted brig. – but I believe an excellent vessel....

Every now & then I have moments of glorious enthusiasm, when I think of the date & cocoa trees, the palms
& ferns so lofty & beautiful – everything new everything
sublime. And if I live to see years in after life how grand
must such recollections be. – Do you know Humboldt? (if
you don't, do so directly) with what intense pleasure he
appears always to look back on the days spent in the tropical countries: I hope, when you next write to Osmaston,
tell them my scheme, & give them my kindest regards &
farewells.

Good bye my dear Fox. Yours ever sincerely | Chas Darwin

To Caroline Darwin   12 November [1831]
                                        November 12th.
My dear Caroline,
    ...Everything here is most prosperous; the Beagle, now
looks something like a ship. – They have just painted her
and in a week's time the men will live on board. – No Vessel
has ever been fitted at all on so expensive a scale from Plymouth. – I get into a fine naval fervour whenever I look at
her. I suppose she is as good a ship as art can make her –
and if I believe all I hear the Captain is as perfect as nature
can make him. – It is ridiculous to see how popular he is,
ladies can hardly splutter out big enough words to express
their big feelings. . . .

NOTES

Darwin's transformation from an amiable, directionless young
man to a genuine scientist who learned how to develop appropriate and effective methodologies (and, later, to establish an
extensive network of useful correspondents) took place aboard
HMS Beagle. Joining the ship as a naturalist, however, required
more than a simple letter of invitation to apply for the position,

written by Henslow. He had to overcome the more serious objections of his father, recruit the support of his uncle Josiah Wedgwood II, and pass the personal inspection of Captain Fitz-Roy. The steps in this process are recapitulated in the letters reproduced here. They begin with George Peacock's letter to the Reverend John Stevens Henslow asking for a recommendation of 'a proper person' (early August 1831), and end with Darwin's high praise of Fitz-Roy as someone 'perfect as nature can make him' (mid-November 1831). Darwin's respect for Fitz-Roy as a natural-born leader of men remained unshaken despite a growing awareness of differences between his temperament and that of the Captain. The tragic clash of philosophical and religious temperaments, which ended only with Fitz-Roy's suicide, came years after the voyage of the *Beagle* had ended.

The reasons why Darwin saw, immediately and clearly, the possibilities of the appointment dangled before him were not immediately understood by most of his contemporaries. His campaign to secure the appointment surprised his father, and after the smoke had cleared, possibly himself. At any rate, the letters are lively with the passion of a young man awakening to a sense of what his true role in future years might entail.

1.   George Peacock (1791–1858) was an Anglican clergyman and astronomer who later became the Lowndean Professor of Astronomy at Cambridge (1836–58) and the Dean of Ely (1839–58). At first Peacock, writing to Henslow, thought that Leonard Jenyns (Henslow's brother-in-law) might be suitable for the post of naturalist aboard the *Beagle*, but nominated, as his second choice, Charles Darwin.

2.   See above p. 53, n 53.

3.   See above p. 60, n 54.

4.   Bum bailiff: a disparaging term used to denote a sheriff's deputy or county court bailiff. The word *bum*, meaning buttocks, implies that the bailiff pursues and catches from behind.

5.   Darwin's father-in-law and uncle.

6.   Darwin, learning that his father disliked the thought of his signing up for the voyage of the *Beagle*, wrote to Peacock,

refusing his offer. Only the efforts of Josiah Wedgwood II, based on a point-by-point analysis of, and a counter-argument against, the multiple objections raised, led Darwin's father to change his mind.

7.   Sir Francis Beaufort (1774–1857), a personal friend of Fitz-Roy, requested George Peacock to make the offer of naturalist to Darwin. Sir Francis was Hydrographer to the Navy.

8.   Sir William Page Wood, Baron Hatherley (1801–81), was a Fellow of Trinity College (1824–79), and later became Lord Chancellor (1868–77). Sir William, though supporting the choice of Darwin as the *Beagle*'s naturalist, warned Fitz-Roy that Darwin was a Whig.

9.   Fitz-Roy's offer of the post of naturalist to a 'friend' was not known either to George Peacock or Sir Francis Beaumont, both of whom acted in good faith when they supported Darwin's candidacy. The identity of the 'friend', referred to as Mr Chester in Darwin's letter to Henslow of 5 September 1831, is not known for certain, though the plausible suggestion has been made that he may have been Harry Chester, a 'valued friend' of Fitz-Roy, and a novelist who, at the time, was working in the Privy Council office.

10.   Another indication that Darwin read and remembered literary texts. The reference is to *Julius Caesar*, Act IV, Scene 3.

11.   Sir Edward Kynaston, Bart. (1775–1839), was Vicar of Kinnerley, Shropshire.

12.   Fitz-Roy eventually rated the qualities of twenty-two chronometers taken on the voyage.

13.   William Yarrell (1784–1856) was a London bookseller, stationer, zoologist and naturalist.

14.   Charles Musters served as one of the 'Volunteers 1st Class' aboard the *Beagle*.

15.   Philip Parker King (1793–1856) was Captain of the *Adventure*, which sailed with the *Beagle* on its first voyage. He was the father of Philip Gidley King (1817–1904), a midshipman aboard the *Beagle* who became a close friend of Darwin.

Robert Fitz-Roy,[1] 'Darwin Helps Save the Lifeboats of *HMS Beagle*', *Narrative of the Surveying Voyages of His Majesty's Ships* Adventure *and* Beagle, *between the Years 1826 and 1836, Describing Their Examination of the Southern Shores of South America, and* The Beagle's *Circumnavigation of the Globe*, vol. II [of III], *Proceedings of the Second Expedition 1831–1836, Under the Command of Captain Fitz-Roy, R. N.* (London: Henry Colburn, 1839; rpt AMS Press, 1966), pp. 18–19, 215–17 [1831–2]

Considering the limited disposable space in so very small a ship, we contrived to carry more instruments and books than one would readily suppose could be stowed away in dry and secure places; and in a part of my own cabin twenty-two chronometers were carefully placed.

Anxious that no opportunity of collecting useful information, during the voyage, should be lost, I proposed to the Hydrographer that some well-educated and scientific person should be sought for who would willingly share such accommodations as I had to offer, in order to profit by the opportunity of visiting distant countries yet little known. Captain Beaufort approved of the suggestion, and wrote to Professor Peacock, of Cambridge, who consulted with a friend, Professor Henslow, and he named Mr Charles Darwin, grandson of Dr Darwin the poet, as a young man of promising ability, extremely fond of geology, and indeed all branches of natural history. In consequence an offer was made to Mr Darwin to be my guest on board, which he accepted conditionally; permission was obtained for his embarkation, and an order given by the Admiralty that he should be borne on the ship's books for provisions. The conditions asked by Mr Darwin were, that he should be at liberty to leave the Beagle and retire from the Expedition when he thought proper, and that he should pay a fair share of the expenses of my table. . . .

• • •

On the 29th we reached Devil Island, and found the large wigwam still standing, which in 1830 my boat's crew called the 'Parliament House'. Never, in any part of Tierra del Fuego, have I noticed the remains of a wigwam which seemed to have been burned or pulled down; probably there is some feeling on the subject, and in consequence the natives allow them to decay naturally, but never wilfully destroy them. We enjoyed a grand view of the lofty mountain, now called Darwin, with its immense glaciers extending far and wide. Whether this mountain is equal to Sarmiento in height, I am not certain, as the measurements obtained did not rest upon satisfactory data; but the result of those measures gave 6,800 feet for its elevation above the sea. This, as an abstract height, is small, but taking into consideration that it rises abruptly from the sea, which washes its base, and that only a short space intervenes between the salt water and the lofty frozen summit, the effect upon an observer's eye is extremely grand, and equal, probably, to that of far higher mountains which are situated at a distance inland, and generally rise from an elevated district.

We stopped to cook and eat our hasty meal upon a low point of land, immediately in front of a noble precipice of solid ice; the cliffy face of a huge glacier, which seemed to cover the side of a mountain, and completely filled a valley several leagues in extent.

Wherever these enormous glaciers were seen, we remarked the most beautiful light blue or sea green tints in portions of the solid ice, caused by varied transmission, or reflection of light. Blue was the prevailing colour, and the contrast which its extremely delicate hue, with the dazzling white of other ice, afforded to the dark green foliage, the almost black precipices, and the deep, indigo blue water, was very remarkable.

Miniature icebergs surrounded us; fragments of the cliff, which from time to time fall into a deep and gloomy basin beneath the precipice, and are floated out into the channel by a slow tidal stream. In the first volume the frequent falling of these masses of ice is noticed by Captain King in the Strait of Magalhaens, and in the narrative of my first exploring visit to this arm of the Beagle Channel; therefore I will add no further remark upon the subject.

Our boats were hauled up out of the water upon the sandy point, and we were sitting round a fire about two hundred yards from them, when a thundering crash shook us – down came the whole front of the icy cliff – and the sea surged up in a vast heap of foam. Reverberating echoes sounded in every direction, from the lofty mountains which hemmed us in; but our whole attention was immediately called to great rolling waves which came so rapidly that there was scarcely time for the most active of our party to run and seize the boats before they were tossed along the beach like empty calabashes. By the exertions of those who grappled them or seized their ropes, they were hauled up again out of reach of a second and third roller; and indeed we had good reason to rejoice that they were just saved in time; for had not Mr Darwin, and two or three of the men, run to them instantly, they would have been swept away from us irrecoverably. Wind and tide would soon have drifted them beyond the distance a man could swim; and then, what prizes they would have been for the Fuegians, even if we had escaped by possessing ourselves of canoes. At the extremity of the sandy point on which we stood, there were many large blocks of stone, which seemed to have been transported from the adjacent mountains, either upon masses of ice, or by the force of waves such as those which we witnessed. Had our boats struck those blocks, instead of soft sand, our dilemma would not have been much less than if they had been at once swept away.

NOTES

Today Robert Fitz-Roy (1805–65) is perhaps most often remembered as the Captain of *HMS Beagle* who engaged Darwin as ship's naturalist. He held views on the sacred and literal reading of the Old Testament so diametrically opposed to those of Charles Darwin that their friendship, weakening in the years following the voyage, was fatally damaged by the publication of *The Origin of Species*. Readers interested in tracing the relationship often fasten on that poignant moment when Fitz-Roy, present during the Huxley–Wilberforce debate, melodramatically declared his allegiance to Holy Scripture (see p. 179 below). Subject to bouts of depression, he finally succumbed to 'a fit of mental aberration' and committed suicide.

He deserves from later generations a more grateful recognition of his accomplishments. His command of the *Beagle* led to the establishment of a chronometric line round the world, facilitating the fixing of the longitude of many secondary meridians; it also earned him the gold medal of the Royal Geographical Society. He held the post of Governor and Commander-in-Chief of New Zealand (1843–5), but, because of his openly-declared sympathy for aborigines who were being cheated on land transactions, became involved in controversies with white settlers, and had to be recalled. The high quality of his scientific work led to his being appointed a 'Meteorological Statist' (1854), i.e. he assumed the directorship of the Meteorological

Department of the Board of Trade. Less than a decade later he published his *Weather Book* (1863), an innovative and wide-ranging discussion of weather problems that immediately became a standard reference (more reliable storm predictions were one consequence of its wide adoption), and pushed forward, with a herculean effort, to advance the cause of the Lifeboat Association.

Fitz-Roy has little to say about Darwin in the *Proceedings*, but he got along very well with Darwin during the long voyage of the *Beagle*, with the exception of one notable moment in Bahia, Brazil. Curious as to whether slaves resented their lack of freedom, Fitz-Roy queried them directly. Later, Darwin noted that he asked Fitz-Roy, 'perhaps with a sneer, whether he thought the answer of slaves in the presence of their master was worth anything'. Fitz-Roy reacted angrily; he forbade Darwin to share his cabin with him; and for a few hours their friendly relationship seemed to have come to a permanent end. (Fitz-Roy finally relented.) His approval of Darwin as the right kind of travelling companion, as well as for the position of naturalist aboard his ship, never wavered.

1. His name is frequently spelled 'Fitzroy' or 'FitzRoy'. However, 'Fitz-Roy' is printed on the title page of the *Narrative*.

Sir Bartholomew James Sulivan, 'Impressions of Charles Darwin' in *Life and Letters of the Late Admiral Sir Bartholomew James Sulivan, K.C.B., 1810–90*, edited by Henry Norton Sulivan (London: John Murray, 1896), pp. 40, 42–3, 46, 381–2 [1831–6]

Sulivan had his share of ordinary boat work, and during Wickham's[1] long absences did first lieutenant's duties. In December 1834 came his turn to undertake a separate survey. On Christmas Eve he started to survey the east side of the Island of Chiloe and the islets in the Gulf of Ancud. With him went Darwin, three officers, and ten men. They returned on January 7th, but went back to the same ground a few days later, Darwin not accompanying them this time. On the 17th they rejoined the *Beagle* in San Carlos Harbour. Some extracts from the accounts of these trips may prove interesting:

RIO NEGRO, September 5th, 1833.

On August 29th I left in the yawl with a mate and ten men. We started from the ship at 1 p.m. with a strong breeze but a favourable tide, and we beat up to Punta Alta in time to have everything landed, the tents rigged, and the pot under way before sunset. Tea is a great luxury in cruises of this kind. We always boiled a large boiler holding four gallons full every morning for breakfast, and the same for supper, and we never had any left, and, as there were only twelve of us, we must have drunk one-eighth of a gallon each meal, or five and a half pints a

day. The same pot full of a mess made of salt pork, fresh
beef, vension, and biscuit was also emptied for dinner, and
meat also of some kind both for breakfast and supper.
Such hardships are hard to put up with, the idea of being
among mud-banks in a boat with nothing but a waterproof awning to cover her with, and thick blankets to
sleep in, with only two pounds of meat, two-thirds of a
gallon of tea, one pound of bread, and a quarter of a pint
of rum each man per day is dreadful!!!

In the evening we got all ready for beginning work at
daylight, and then part went on board the boat to sleep.
On the 30th we began at 6 a.m., and had finished our
work by breakfast-time, but waited for Darwin to examine the beach at low water for fossil remains of animals,
which are very plentiful. Besides getting some he had
seen before, he this morning found the teeth of animals
six times as large as those of any animal now known in this
country, also the head of one about the size of a horse,
with the teeth quite perfect and totally different from any
now known, and just at low-water mark he found the
remains of another about six feet long, nearly perfect, all
embedded in solid rock. We started at low water for the
settlement, leaving two hands digging out the bones.

After supper we all went on board, and moored the
boat head and stern about four yards from the bushes, to
ensure her grounding in the centre where the mud was
quite soft. The evening looked very gloomy, with heavy
thunder and lightning; but we were quite snug under an
awning, which we filled as much as possible with tobacco
smoke, to drive away the mosquitoes and sand-flies,
which were very troublesome. By filling the upper part of
the awning with smoke we kept them all out. I never in
my life, I think, laughed in the way I did for about three
hours at the stories they were all telling in turns. We had
among the men two or three excellent hands for keeping
every one alive, and to-night they performed their part to
perfection. Such hands are invaluable in a cruise of that
kind, particularly if the work is very hard, as they keep
men's spirits up in a most surprising manner. I think

I never in my life saw people more happy than all our party were; they were in roars of laughter from morning till night, and up to all kinds of amusements when on shore, except when I brought them to an anchor occasionally to prevent their shaking the ground (near my instruments), and then they would find something amusing in that; and when men in those spirits are happy and comfortable, it is astonishing how they make work fly.

* * *

[There follows a short extract from a letter by Sir Thomas Henry Farrer, Bart., 1st Baron Farrer (1819–99), who wrote a long (undated) letter, containing his reminiscences of Sir Bartholomew James Sulivan and his relationship to Charles Darwin, to Henry Norton Sulivan, who at the time was collecting material for his biography – Ed.]

I might add, in relation to Darwin, that he suffered so much from seasickness that whenever the ship was out of harbour he retired to his hammock in the chart-room, the only accommodation afforded him. My father said he believed it was this constant suffering which laid the seeds of the indisposition he was troubled with in later years, and that his patience in persevering with his scientific work, and not abandoning the voyage, was most commendable.

• • •

I used to hear much of the voyage of the *Beagle* from Charles Darwin, whose niece I married, and whose son has married my daughter. He and the other partners in that historical cruise had an infinite respect for FitzRoy, whose abilities as a seaman, whose courage as a man, and whose self-sacrifice as a public servant, won the esteem and admiration, if they did not command the affection, of all who served under him. From all I heard, all the officers and men loved Sulivan; certainly Charles Darwin did.

There must have been something very good and strong about those men to keep them together for so many years, cooped up in that small uncomfortable vessel, doing first-rate work with very inadequate means.

> One equal temper of heroic hearts,
> Not strong in circumstance, but strong in will.
> To seek, to find, to strive, and not to yield.[2]

NOTES

Sir Bartholomew James Sulivan (1810–90) participated in important voyages that alternated surveying and combat duties (Africa, the Falkland Islands during the hostilities of 1845, and the Baltic during the campaigns of 1854 and 1855). He assumed major responsibility for developing the concept of a naval reserve while assigned to *HMS Victory* (1847–8).

One of his early important postings was to *HMS Beagle*. Fitz-Roy, serving with Sulivan aboard *HMS Thetis*, was appointed Captain of the *Beagle* in 1828, and immediately requested that Sulivan – who admired him greatly – be transferred to his new command as a Second Lieutenant; thus Sulivan served not only on the first extended voyage but, along with Darwin, on the more famous second voyage.

The above excerpts testify to Sulivan's respect for a landlubber who performed the duties of a sea-going naturalist in a thoroughly professional manner.

1.   John Clements Wickham (1798–1864), a First Lieutenant and the oldest officer on board (he was 33), served on all three voyages of the *Beagle* (he commanded the third). Darwin enjoyed his company more than that of any other member of the crew. When Fitz-Roy, in a severely depressed mood, briefly gave up his command (September 1834), Wickham served as a replacement Captain.

2.   Alfred, Lord Tennyson, 'Ulysses'.

'Darwin as a Welcome Guest', in *Harriet Martineau's Autobiography*, edited by Maria Weston Chapman (Boston: James R. Osgood and Company, 1877), vol. I, p. 268 [1836–40s]

In what noble contrast were the eminent men who were not vain! There was the honest and kindly Captain (now Admiral Sir Francis) Beaufort,[1] who was daily at the Admiralty as the clock struck, conveying paper, pen and ink for any private letters he might have to write, for which he refused to use the official stores. There were the friends Lyell and Charles Darwin – after the return of the latter from his four years' voyage round the world; – Lyell with a Scotch prudence which gave way, more and more as years passed on, to his natural geniality, and to an expanding liberality of opinion and freedom of speech; and the simple, childlike, painstaking, effective Charles Darwin, who established himself presently at the head of living English naturalists. These well-employed, earnest-minded, accomplished and genial men bore their honours without vanity, jealousy, or any apparent self-regard whatever. They and their devoted wives were welcome in the highest degree....

NOTES

For a note on Harriet Martineau, see below, pp. 162–3.

1.   Captain Beaufort: see above, p. 77, n. 7.

'Owen Studies a Darwin Specimen', in *The Life of Richard Owen*, by Reverend Richard Owen (London: John Murray, 1894), vol. I, pp. 119–21 [1838]

In 1838 Owen wrote a paper, which was the nucleus of his great work on teeth – the 'Odontography'. This paper was entitled: 'On the Structure of Teeth, and the Resemblance of Ivory to Bone, as illustrated by the Microscopical Examination of the Teeth of Men and of various Existing and Extinct Animals'. ('Report of the British Association, 1838'.)

Amongst the descriptions which Owen made of the fossil mammalia collected by Darwin in the voyage of the 'Beagle' may be mentioned that of the *Toxodon* skull. The toxodon was a gigantic extinct mammal, presenting great peculiarities and having points in common with various orders of Mammalia.

The following account of the toxodon in the autograph of Charles Darwin was found amongst Owen's papers, from which an extract is now given: –

'The head was found embedded in whitish earthy clay on the banks of a small stream which enters the Rio Negro, and is situated 120 miles to the N. W. of Monte Video. The head had been kept for a short time in a neighbouring farm-house as a curiosity, but when I arrived it was lying in the yard. I bought it for the value of eighteen-pence. The people informed me that when first discovered, about two years previously, it was quite perfect, but that the boys had since knocked out the teeth and had put it on a post as a mark to throw stones at. They showed me the spot where it had been found after a sudden flood had washed down part of the bank. Several fragments of bone and of an armadillo-like case were lying at the bottom of the almost dry watercourse. Some of these I collected, but from the disturbed state of the country the box in which they were packed was delayed on the road, and was afterwards sent direct to England.

'For this reason the temporary marks by which I had distinguished these bones from another set, found at the distance of several leagues, were lost, and I am now unable to say which are the fragments.... This river (Rio Cancaraña) has been celebrated since the time of the Jesuit Falkner for the number of great bones and large fragments of the armadillo-like case found in its bed. The inhabitants told me that they had made gate-posts of some leg bones, and I myself saw two groups *in situ* of the remains of a mastodon projecting from a cliff. But they were in so decayed a state that I could only bring away small portions of a molar tooth.'

From the same collection Owen described the remains of an extinct animal related to the llama. He also described the scelidotherium, which is related to the ant-eaters; and further determined some disputed points in existing accounts of the skeleton of the megatherium – a gigantic extinct sloth about the size of an elephant. We also find from the Diary that Darwin submitted the proofs of the 'Voyage' itself to Owen.

But while occupied in describing fossil remains he varied his occupation by dissecting the mortal remains of a rhinoceros which had recently died at Wombwell's Menagerie.[1] This he looked upon as a great prize, as a rhinoceros then – dead or living – was a rarity in England. On February 1, Owen had the carcase brought to his house in the College of Surgeons, to his wife's disgust, who thus comments upon it: – 'The defunct rhinoceros (late of Wombwell's Menagerie) arrived while R. was out. I told the men to take it right to the end of the long passage, where it now lies. As yet I feel indifferent, but when the pie is opened ——'.

'*February* 6. – R. still at the rhinoceros.'

NOTES

Sir Richard Owen (1804–92), an enormously energetic and productive biologist, created – probably more than any single other individual – the dream, and shaped the reality, of a National Museum of Natural History in South Kensington. Owen's pioneering work in preparing catalogues on the Hunterian Collection in the Royal College of Surgeons (where he held the post of Conservator), and his extensive research on invertebrate animals and Vertebrata, earned him recognition as England's foremost biologist.

It would be pleasant to report that he encouraged Darwin's research in a consistent, constructive fashion. He certainly understood what Darwin's findings implied for the future development of England's science. Yet he wrote an anonymous, and hostile, review of *The Origin of Species* for the *Edinburgh Review* (April 1860), and may well have inspired the Bishop of Oxford's article in the *Quarterly Review* in the same month (Darwin believed Owen was responsible for the article). Darwin, who soon learned the identity of his 'enemy' (a man whose accomplishments he had always appreciated, even admired), knew that he disagreed with Owen on fundamental matters pertaining to organic evolution, but he despised Owen's underhanded tactics, and became angrier with Owen's opposition than with that raised by any of his other critics. (As his contemporaries noted, Darwin's cheerful willingness to engage fellow scientists in serious debate precluded practically all opportunities for losing his temper about a difference of opinion with a single individual.) Matters were not helped by Owen's increasing dependence on a deliberately ambiguous style (he was already notorious for his difficult prose), but perhaps that was simply an indication that he rested content with his own vague formulations of biological philosophy. These opinions were to become increasingly unacceptable to younger generations.

Owen's grandson, the Reverend Richard Owen, wrote as sympathetically as he could about the sharp-tempered dispute between two of the foremost researchers of the age; but it is clear that Darwin had ample reason to feel aggrieved.

1. Wombwell's Royal No. 1 Menagerie attracted thousands of visitors between 1848 and 1871. In 1855 a queer sort of chimpanzee (so characterized by the owners of the Menagerie) was identified by a zoologist as a female gorilla, the first one ever to be in captivity outside the Congo and Cameroon forests. The collections of the Menagerie were dispersed in Edinburgh in 1872. See James Fisher, *Zoos of the World / The Story of Animals in Captivity* (Garden City, NY: Natural History Press, in the Nature and Science Library, published by the American Museum of Natural History, 1967), p. 89.

Charles Darwin, 'The Death of Anne Elizabeth Darwin', in *The Correspondence of Charles Darwin*, vol. 5: *1851–55* (Cambridge: Cambridge University Press, 1985), Appendix II, pp. 540–2 [1851]

Our poor child, Annie, was born in Gower St on March 2d. 1841. & expired at Malvern at Midday on the 23d. of April 1851. – I write these few pages, as I think in after years, if we live, the impressions now put down will recall more vividly her chief characteristics. From whatever point I look back at her, the main feature in her disposition which at once rises before me is her buoyant joyousness tempered by two other characteristics, namely her sensitiveness, which might easily have been overlooked by a stranger & her strong affection. Her joyousness and animal spirits radiated from her whole countenance & rendered every movement elastic & full of life & vigour. It was delightful & cheerful to behold

her. Her dear face now rises before me, as she used sometimes to come running down stairs with a stolen pinch of
snuff for me, her whole form radiant with the pleasure of
giving pleasure. Even when playing with her cousins when
her joyousness almost passed into boisterousness, a single
glance of my eye, not of displeasure (for I thank God I
hardly ever cast one on her), but of want of sympathy
would for some minutes alter her whole countenance. This
sensitiveness to the least blame, made her most easy to
manage & very good: she hardly ever required to be found
fault with, & was never punished in any way whatever. Her
sensitiveness appeared extremely early in life, & showed
itself in crying bitterly over any story at all melancholy; or
on parting with Emma even for the shortest interval. Once
when she was very young she exclaimed 'Oh Mamma,
what should we do, if you were to die.'

The other point in her character, which made her joyousness & spirits so delightful, was her strong affection, which
was of a most clinging, fondling nature. When quite a Baby,
this showed itself in never being easy without touching
Emma, when in bed with her, & quite lately she would
when poorly fondle for any length of time one of Emma's
arms. When very unwell, Emma lying down beside her,
seemed to soothe her in a manner quite different from what
it would have done to any of our other children. So again,
she would at almost anytime spend half-an-hour in arran-
ging my hair, 'making it' as she called it 'beautiful', or in
smoothing, the poor dear darling, my collar or cuffs, in
short in fondling me. She liked being kissed; indeed every
expression in her countenance beamed with affection &
kindness, & all her habits were influenced by her loving disposition.

Besides her joyousness thus tempered, she was in her
manners remarkably cordial, frank, open, straightforward
natural and without any shade of reserve. Her whole mind
was pure & transparent. One felt one knew her thoroughly
& could trust her: I always thought, that come what might,
we should have had in our old age, at least one loving soul,

which nothing could have changed. She was generous, handsome & unsuspicious in all her conduct; free from envy & jealousy; goodtempered & never passionate. Hence she was very popular in the whole household, and strangers liked her & soon appreciated her. The very manner in which she shook hands with acquaintances showed her cordiality.

Her figure & appearance were clearly influenced by her character: her eyes sparkled brightly; she often smiled; her step was elastic & firm; she held herself upright, & often threw her head a little backwards, as if she defied the world in her joyousness. For her age she was very tall, not thin & strong. Her hair was a nice brown & long; her complexion slightly brown; eyes, dark grey; her teeth large & white. The Daguerrotype is very like her, but fails entirely in expression: having been made two years since, her face had become lengthened & better looking. All her movements were vigorous, active & usually graceful: when going round the sand-walk with me, although I walked fast, yet she often used to go before pirouetting in the most elegant way, her dear face bright all the time, with the sweetest smiles.

Occasionally she had a pretty coquettish manner towards me; the memory of which is charming: she often used exaggerated language, & when I quizzed her by exaggerating what she had said, how clearly can I now see the little toss of the head & exclamation of 'Oh Papa what a shame of you'. – She had a truly feminine interest in dress, & was always neat: such undisguised satisfaction, escaping somehow all tinge of conceit & vanity, beamed from her face, when she had got hold of some ribbon or gay handkerchief of her Mamma's. – One day she dressed herself up in a silk gown, cap, shawl & gloves of Emma, appearing in figure like a little old woman, but with her heightened colour, sparkling eyes & bridled smiles, she looked, as I thought, quite charming.

She cordially admired the younger children; how often have I heard her emphatically declare 'what a little duck, Betty[1] is, is not she?'

She was very handy, doing everything neatly with her hands: she learnt music readily, & I am sure from watching her countenance, when listening to others playing, that she had a strong taste for it. She had some turn for drawing, & could copy faces very nicely. She danced well, & was extremely fond of it. She liked reading, but evinced no particular line of taste. She had one singular habit, which, I presume would ultimately have turned into some pursuit; namely a strong pleasure in looking out words or names in dictionaries, directories, gazeteers, & in this latter case finding out the places in the Map: so also she would take a strange interest in comparing word by word two editions of the same book; and again she would spend hours in comparing the colours of any objects with a book of mine, in which all colours are arranged & named.

Her health failed in a slight degree for about nine months before her last illness; but it only occasionally gave her a day of discomfort: at such times, she was never in the least degree cross, peevish or impatient; & it was wonderful to see, as the discomfort passed, how quickly her elastic spirits brought back her joyousness & happiness. In the last short illness, her conduct in simple truth was angelic; she never once complained; never became fretful; was ever considerate of others; & was thankful in the most gentle, pathetic manner for everything done for her. When so exhausted that she could hardly speak, she praised everything that was given her, & said some tea 'was beautifully good'. When I gave her some water, she said 'I quite thank you'; & these, I believe, were the last precious words ever addressed by her dear lips to me.

But looking back, always the spirit of joyousness rises before me as her emblem and characteristic: she seemed formed to live a life of happiness: her spirits were always held in check by her sensitiveness lest she should displease those she loved, & her tender love was never weary of displaying itself by fondling & all the other little acts of affection.

We have lost the joy of the Household, and the solace of our old age: – she must have known how we loved her; oh that she could now know how deeply, how tenderly we do still & shall ever love her dear joyous face. Blessings on her.

April 30. 1851.

NOTES

This emotion-laden memoir was written in 1851, a few days after the death of Darwin's beloved eldest daughter, Anne Elizabeth Darwin. She had never fully recovered from an attack of scarlet fever (1849), and Darwin believed that she had inherited from him a 'wretched digestion' and a 'hereditary weakness'. Nevertheless, he was hopeful that Dr James Manby Gully (1808–83), a hydropathic specialist, might be able, by means of a water treatment, to restore her to full health. The blow of her passing, compounded by the fact that Emma's pregnancy had prevented her from being with Anne during the final stage of her illness, was very severe. He called Anne his 'dear and good child', one who was 'in simple truth angelic'; but he could no longer pretend to a faith in a traditional God, or in an afterlife. His opinions were to become, in the 1850s, a source of deep, growing unhappiness to his wife, a woman who based her religion on what the Bible told her was divine or revealed truth. The linkage between that fact and the fact that his anguish, continuing over a lifetime, prevented him from visiting Malvern Churchyard, where Anne was buried, seems indisputable.

1.   Elizabeth Darwin ('Betty') was six years younger than Anne.

Alfred Russel Wallace, 'A Developing Friendship with Darwin', in *My Life / A Record of Events and Opinions* (New York: Dodd, Mead & Company, 1905), vol. I, pp. 360–3; vol. II, pp. 1–16 [1858–81]

It was while waiting at Ternate in order to get ready for my next journey, and to decide where I should go, that the idea already referred to occurred to me. It has been shown how, for the preceding eight or nine years, the great problem of the origin of species had been continually pondered over, and how my varied observations and study had been made use of to lay the foundation for its full discussion and elucidation. My paper written at Sarawak rendered it certain to my mind that the change had taken place by natural succession and descent – one species becoming changed either slowly or rapidly into another. But the exact process of the change and the causes which led to it were absolutely unknown and appeared almost inconceivable. The great difficulty was to understand how, if one species was gradually changed into another, there continued to be so many quite distinct species, so many which differed from their nearest allies by slight yet perfectly definite and constant characters. One would expect that if it was a law of nature that species were continually changing so as to become in time new and distinct species, the world would be full of an inextricable mixture of various slightly different forms, so that the well-defined and constant species we see would not exist. Again, not only are species, as a rule, separated from each other by distinct external characters, but they almost always differ also to some degree in their food, in the places they frequent, in their habits and instincts, and all these characters are quite as definite and constant as are the external characters. The problem then was, not only how and why do species change, but how and why do they change into new and well-defined species, distinguished from each other in so many ways; why and how do they become so exactly adapted to distinct modes of life; and why do all the intermediate grades die out (as geology shows they have died out) and leave only clearly defined and well-marked species, genera, and higher groups of animals.

Now, the new idea or principle which Darwin had arrived at twenty years before, and which occurred to me at this time, answers all these questions and solves all these difficulties, and it is because it does so, and also because it is in itself self-evident and absolutely certain, that it has been accepted by the whole scientific world as affording a true solution of the great problem of the origin of species.

At the time in question I was suffering from a sharp attack of intermittent fever, and every day during the cold and succeeding hot fits had to lie down for several hours, during which time I had nothing to do but to think over any subjects then particularly interesting me. One day something brought to my recollection Malthus's 'Principles of Population', which I had read about twelve years before. I thought of his clear exposition of 'the positive checks to increase' – disease, accidents, war, and famine – which keep down the population of savage races to so much lower an average than that of more civilized peoples. It then occurred to me that these causes or their equivalents are continually acting in the case of animals also; and as animals usually breed much more rapidly than does mankind, the destruction every year from these causes must be enormous in order to keep down the numbers of each species, since they evidently do not increase regularly from year to year, as otherwise the world would long ago have been densely crowded with those that breed most quickly. Vaguely thinking over the enormous and constant destruction which this implied, it occurred to me to ask the question, Why do some die and some live? And the answer was clearly, that on the whole the best fitted live. From the effects of disease the most healthy escaped; from enemies, the strongest, the swiftest, or the most cunning; from famine, the best hunters or those with the best digestion; and so on. Then it suddenly flashed upon me that this self-acting process would necessarily *improve the race*, because in every generation the inferior would inevitably be killed off and the superior would remain – that is, *the fittest would survive*. Then at once I seemed to see the whole effect of this, that when changes of land and sea, or of climate, or of food-supply, or of enemies occurred – and we know that such changes have always been taking place – and considering the amount of individual variation that my experience as a collector had shown me to exist, then it followed that all the changes necessary for the adaptation of the species to the changing conditions would be brought about; and as great changes in the environment are always slow, there would be ample time for the change to be effected by the survival of the best fitted in every generation. In this way every part of an animal's organization could be modified exactly as required, and in the very process of this modification the unmodified would die out, and thus the *definite* characters and the clear *isolation* of each new species would be explained. The more I thought over it the more I became convinced that I had at length found the long-sought-for law of nature that solved the problem of the origin of species. For the next hour I thought over the deficiencies in the theories of Lamarck and of the author of the 'Vestiges', and I saw that my new theory supplemented these views and obviated every important difficulty. I waited anxiously for the termination of my fit so that I might at once make notes for a paper on the subject. The same evening I did this pretty fully, and on the two succeeding evenings wrote it out carefully in order to send it to Darwin by the next post, which would leave in a day or two.

I wrote a letter to him in which I said that I hoped the idea would be as new to him as it was to me, and that it would supply the missing factor to explain the origin of species. I asked him if he thought it sufficiently important to show to Sir Charles Lyell, who had thought so highly of my former paper.

The subsequent history of this article is fully given in the 'Life and Letters', volume ii., and I was, of course, very much surprised to find that the same idea had occurred to Darwin, and that he had already nearly completed a large work fully developing it. The paper is reprinted in my 'Natural Selection and Tropical Nature', and in reading it *now* it must be remembered that it was but a hasty first sketch, that I had no opportunity of revising it before it was printed in the journal of the Linnaean Society, and, especially, that at that time nobody had any idea of the constant variability of *every* common species, in every part and organ, which has since been proved to exist. Almost all the popular objections to Natural Selections are due to ignorance of this fact, and to the erroneous assumption that what are called 'favourable variations' occur only rarely, instead of being abundant, as they certainly are, in every generation, and quite large enough for the efficient action of 'survival of the fittest' in the improvement of the race.

•　•　•

Soon after I returned home, in the summer of 1862, Mr Darwin invited me to come to Down for a night, where I had the great pleasure of seeing him in his quiet home, and in the midst of his family. A year or two later I spent a weekend with him in company with Bates,[1] Jenner Weir,[2] and a few other naturalists; but my most frequent interviews with him were when he spent a few weeks with his brother, Dr Erasmus Darwin, in Queen Anne Street, which he usually did every year when he was well enough, in order to see his friends and collect information for his various works. On these occasions I usually lunched with him and his brother, and sometimes one other visitor, and had a little talk on some of the matters specially interesting him. He also sometimes called on me in St Mark's Crescent for a quiet talk or to see some of my collections.

My first letter from him dealing with scientific matters was in August, 1862, and our correspondence was very extensive during the period occupied in writing or correcting his earlier books on evolution, down to the publication of 'The Expression of the Emotions in Man and Animals', in 1872, and afterwards, at longer intervals, to within less than a year of his death. A considerable selection of our correspondence has been published in the 'Life and Letters' (1887), and especially in 'More Letters' (1903); while several of the more interesting of these were contained in the one-volume life, entitled 'Charles Darwin', which appeared in 1892. As many of my readers, however, may not have these works to refer to, I will here give a few of his letters to myself which have not yet been published, together with some of my own, and also occasional extracts from some of Darwin's that have already appeared, in order to make clear the nature of our discussions, and also, perhaps, to throw a little light upon our respective characters.

In a letter entirely without date, but which was evidently written in 1863, he gives me some information for which I had asked about reviews of the 'Origin of Species'.

Down, Bromley, Kent (1863).

MY DEAR MR. WALLACE,

I write one line to thank you for your note, and to say that the B. of Oxford wrote the *Quarterly R.* (paid £60), aided by Owen. In the *Edinburgh*, Owen no doubt praised himself. Mr Maw's review in *Zoologist* is one of the best, and staggered me in parts, for I did not see the sophistry of (those) parts. I could lend you any which you might wish to see, but you would soon be tired. Hopkins in *Fraser* and Pictet are two of the best.

I am glad you like the little orchid book; but it has not been worth the ten months it has cost me; it was a hobby horse, and so beguiled me.

How puzzled you must be to know what to begin at! You will do grand work, I do not doubt. My health is, and always will be, very poor; I am that miserable animal, a regular valetudinarian.

Yours very sincerely,

C. DARWIN.

In March, 1864, he wrote me from Malvern Wells that he had been very ill at home, having fits of vomiting every day for two months, and been able to do nothing. These attacks were brought on by the least mental excitement, which often rendered it impossible for him to see his friends, and which appear to have lasted at intervals throughout his life. This must always be remembered when we consider the enormous amount of work he was able to do; but, unfortunately, the quiet interest of carrying out observations or experiments lasting for months, and often for years, seem to have been beneficial. On the other hand, writing his books and correcting the MSS. and the proofs in the very careful manner he always practised was most wearying and distasteful to him.

On February 23, 1867, he wrote to me asking if I could solve a difficulty for him. He says; 'On Monday evening I called on Bates, and put a difficulty before him which he could not answer, and, as on some similar occasion, his first suggestion was, "You had better ask Wallace." My difficulty is, Why are caterpillars sometimes so beautifully and artistically coloured? Seeing that many are coloured to escape dangers, I can hardly attribute their bright colour in other cases to mere physical conditions. Bates says the most gaudy caterpillar he ever saw in Amazonia was conspicuous at the distance of yards, from its black and red colours, whilst feeding on large, green leaves. If anyone objected to male butterflies having been made beautiful by sexual selection, and asked why they should not have been made beautiful as well as their caterpillars, what would you answer? I could not answer, but should maintain my ground. Will you think over this, and some time, either by letter or when we meet, tell me what you think?'

On reading this letter, I almost at once saw what seemed to be a very easy and probable explanation of the facts. I had then just been preparing for publication (in the *Westminster Review*) my rather elaborate paper on 'Mimicry and Protective Colouring', and the numerous cases in which specially showy and slow-flying butterflies were known to have a peculiar odour and taste which protected them from the attacks of insect-eating birds and other animals, led me at once to suppose that the gaudily-coloured caterpillars must have a similar protection. I had just ascertained from Mr Jenner Weir that one of our common white moths (*Spilosoma menthrastri*) would not be eaten by most of the small birds in his aviary, nor by young turkeys. Now, as a *white* moth is as conspicuous in the *dusk* as a *coloured* caterpillar in the *daylight*, this case seemed to me so much on a par with the other that I felt almost sure my explanation would turn out correct. I at once wrote to Mr Darwin to this effect, and his reply, dated February 26, is as follows:

My Dear Wallace

Bates was quite right; you are the man to apply to in a difficulty. I never heard anything more ingenious than your suggestion, and I hope you may be able to prove it true. That is a splendid fact about the white moths; it warms one's very blood to see a theory thus almost proved to be true.

The following week I brought the subject to the notice of the Fellows of the Entomological Society at their evening meeting (March 4), requesting that any of them who had the opportunity would make observations or experiments during the summer in accordance with Mr Darwin's suggestion. I also wrote a letter to *The Field* newspaper, which, as it explains my hypothesis in simple language, I here give entire:

Caterpillars and Birds

Sir,

May I be permitted to ask the co-operation of your readers in making some observations during the coming spring and summer which are of great interest to Mr Darwin and myself? I will first state what observations are wanted and then explain briefly why they are wanted. A number of our smaller birds devour quantities of caterpillars, but there is reason to suspect that they do not eat all alike. Now we want direct evidence as to which species they eat and which they reject. This may be obtained in two ways. Those who keep insectivorous birds, such as thrushes, robins, or any of the warblers (or any other that will eat caterpillars), may offer them all the kinds they can obtain, and carefully note (1) which they eat, (2) which they refuse to touch, and (3) which they seize but reject. If the name of the caterpillar cannot be ascertained, a short description of its more prominent characters will do very well, such as whether it is hairy or smooth, and what are its chief colours, especially distinguishing such as are green or brown from such as are of bright and conspicuous colours, as yellow, red, or black. The food plant of the caterpillar should also be stated when known. Those who do not keep birds, but have a garden much frequented by birds, may put all the caterpillars they can find in a soup plate or other vessel, which must be placed in a larger vessel of water, so that the creatures cannot escape, and then after a few hours note which have been taken and which left. If the vessel could be placed where it might be watched from a window, so that the kind of birds which took them could also be noted, the experiment would be still more complete. A third set of observations might be made on young fowls, turkeys, guinea-fowls, pheasants, etc., in exactly the same manner.

Now the purport of these observations is to ascertain the law which had determined the coloration of caterpillars. The analogy of many other insects leads us to believe that all those which are green or brown, or of such speckled or mottled tints as to resemble closely the leaf or bark of the plant on which they feed, or the substance on which they usually repose, are thus to some degree protected from the attacks of birds and other enemies. We should expect, therefore, that all which are thus protected would be greedily eaten by birds whenever they can find them. But there are other caterpillars which seem coloured on purpose to be conspicuous, and it is very important to know whether they have another kind of protection, altogether independent of disguise, such as a disagreeable odour and taste. If they are thus protected, so that the majority of birds will never eat them, we can understand that to get the full benefit of this protection they should be easily recognized, should have some outward character by which birds would soon learn to know them and thus let them alone; because if birds could not tell the eatable from the uneatable till they had seized and tasted them, the protection would be of no avail, a growing caterpillar being so delicate that a wound is certain death. If, therefore, the eatable caterpillars derive a partial protection from their obscure and imitative colouring, then we can understand that it would be an advantage to the uneatable kinds to be well distinguished from them by bright and conspicuous colours.

I may add that this question has an important bearing on the whole theory of the origin of the colours of animals, and especially of insects. I hope many of your readers may be thereby induced to make such observations as I have indicated, and if they will kindly send me their notes at the end of the summer, or earlier, I will undertake to compare and tabulate the whole, and to make known the results, whether they confirm or refute the theory here indicated.

ALFRED R. WALLACE.

9 St Mark's Crescent, Regent's Park, N.W.,
March, 1867.

This letter brought me only one reply, from a gentleman in Cumberland, who informed me that the common 'gooseberry' caterpillar, which is the larva of the magpie moth (*Abraxus grossulariata*), is refused by young pheasants, partridges, and wild ducks, as well as sparrows and finches, and that all birds to whom he offered it rejected it with evident dread and abhorrence. But in 1869 two entomologists, Mr Jenner Weir and Mr A. G. Butler,[3] gave an account of their two seasons' experiments and observations with several of our most gaily-coloured caterpillars, and with a considerable variety of birds, and also with lizards, frogs, and spiders, confirming my explanation in a most remarkable manner. An account of these experiments is given in the second and all later editions of my book on 'Natural Selection'; but it is more fully treated in my 'Darwinism', chap. ix, under the heading 'Warning Colours among Insects', and it has thus led to the establishment of a general principle which is very widely applicable, and serves to explain a not inconsiderable proportion of the colours and markings in the animal world. It is, of course, only a wider application of the same fundamental fact by which Bates had already explained the purpose of 'mimicry' among insects, and it is a matter of surprise to me that neither Bates himself nor Darwin had seen the probability of the occurrence of inedibility in the larvae as well as in the perfect insects.

In the year 1870 Mr A. W. Bennett[4] read a paper before Section D of the British Association at Liverpool, entitled, 'The Theory of Natural Selection from a Mathematical Point of View', and this paper was printed in full in *Nature* of November 10, 1870. To this I replied on November 17, and my reply so pleased Mr Darwin that he at once wrote to me as follows:

Down, November 22.

MY DEAR WALLACE,

I must ease myself by writing a few words to say how much I and all in this house admire your article in *Nature*. You are certainly an unparalleled master in lucidly stating a case and in arguing. Nothing ever was better done than your argument about the term Origin of Species, and about much being gained if we know nothing about precise cause of each variation.

[At the end of the letter he says something about the progress of his great work, 'The Descent of Man'. – ARW]

"I have finished 1st vol. and am half-way through proofs of 2nd vol. of my confounded book, which half kills me by fatigue, and which I fear will quite kill me in your good estimation.

"If you have leisure, I should much like a little news of you and your doings and your family.

"Ever yours very sincerely,

CH. DARWIN.

The above remark, 'kill me in your good estimation', refers to his views on the mental and moral nature of man being very different from mine, this being the first important question as to which our views had diverged. But I never had the slightest feeling of the kind he supposed, looking upon the difference as one which did not at all affect our general agreement, and also as being one on which no one could dogmatize, there being much to be said on both sides. The last paragraph shows the extreme interest he took in the personal affairs of all his friends.

As my article of which he thought so highly is buried in an early volume of *Nature*, I will here reproduce the rather long paragraph which so specially interested him. It is as follows:

'The first objection brought forward (and which had been already advanced by the Duke of Argyll) is, that the very title of Mr Darwin's celebrated work is a misnomer, and that the real "origin of species" is that spontaneous tendency to variation which has not yet been accounted for. Mr Bennett further remarks that, throughout my volume of "Essays", I appear to be unconscious that the theory I advocate does not go to the root of the matter. It is true that I am "unconscious" of anything of the kind, for I maintain, and am prepared to prove, that the theory, if true, does go to the very root of the question of the "origin of species". The objection, which from its being so often made, and now again brought forward, is evidently thought to be an important one, is founded on a misapprehension of the right meaning of words. It ignores the fact that the word "species" denotes something more than "variety" or "individual". A species is an organic form (or group) which, for periods of great and indefinite length, as compared with the duration of human life, fluctuates only within narrow limits. But the "spontaneous tendency to variation" is altogether antagonistic to such comparative stability, and would, if unchecked, entirely destroy all "species". Abolish, if possible, selection and survival of the fittest, so that every spontaneous variation should survive in equal proportion with all others, and the result must inevitably be an endless variety of *unstable forms*, no one of which would answer to what we mean by the word "species". No other cause but selection has yet been discovered capable of perpetuating and giving stability to some forms, and causing the disappearance of others, and therefore Mr Darwin's book, if there is any truth in it at all, has a logical claim to its title. It shows how "species", or stable forms, are produced out of unstable spontaneous variations, which is certainly to trace their "origin". The distinction of "species" and "individual" is equally important. A horse, or a number of horses, as such, do not constitute a "species". It is the comparative *permanence* of the form as distinguished from the ass, quagga, zebra, tapir, camel, etc., that makes them one. Were there a mass of intermediate forms connecting all these animals by fine gradations and hardly a dozen individuals alike – as would probably be the case had selection not acted – there might be a few horses, but there would be no such thing as a *species* of horse. That could only be produced by some power capable of eliminating intermediate forms as they arose, and preserving all of the true horse type; and such a power was first shown to exist by Mr Darwin. The origin of varieties and individuals is one thing, the origin of species another'.

It is a remarkable thing that this very simple preliminary misunderstanding of the very meaning of the term 'species' continued to appear year after year in most of the criticisms of the theory of natural selection. It was put forward both by mere literary critics and also by naturalists, and was in many cases adduced as a discovery which completely overthrew the whole of Darwin's work. So frequent was it that twenty years later, when writing my 'Darwinism' I found it necessary to devote the first chapter to a thorough explanation of this point, under the heading, 'What are "Species", and what is meant by their "Origin"?' and I think I may feel confident that to those who have read that work this particular purely imaginary difficulty will no longer exist.

Soon after the 'Descent of Man' appeared, I wrote to Darwin, giving my impressions of the first volume, to which he replied (January 30, 1871). This letter is given in the 'Life and Letters' (iii. p. 134), but I will quote two short passages expressing his kind feelings towards myself. He begins, 'Your note has given me very great pleasure, chiefly because I was so anxious not to treat you with the least disrespect, and it is so difficult to speak fairly when differing from anyone. If I had offended you, it would have grieved me more than you will readily believe.' And the conclusion is, 'Forgive me for scribbling at such length. You have put me quite in good spirits; I did so dread having been unintentionally unfair towards your views. I hope earnestly the second volume will escape as well. I care now very little what others say. As for our not agreeing, really, in such complex subjects, it is almost impossible for two men who arrive independently at their conclusions to agree fully; it would be unnatural for them to do so.'

I reviewed 'The Descent', in *The Academy*, early in March, and Darwin wrote to me on the 16th, expressing his gratification at its whole tone and matter, and then, referring to the differences between us, making what was then a good point against me – that my objections to sexual selection having produced certain results in man, had not much force if, as he believed, I admitted that the plumes of the birds of paradise had thus been gained. At that time, though I had begun to doubt, I had not definitely rejected the whole of

that part of 'sexual selection' depending on female preference for certain colours and ornaments.

On July 9, 1871, he wrote me a long letter, chiefly about
Mr Mivart's[5] criticisms and accusations in his book on 'The
Genesis of Species', and again in a severe article in the *Quarterly Review*. These he proposed replying to in a new edition
of the 'Origin,' but the incident worried him a good deal. In
a postscript he says, 'I quite agree with what you say, that
Mivart fully intends to be honourable, but he seems to me
to have the mind of a most able lawyer retained to plead
against us, and especially against me. God knows whether
my strength and spirit will last out to write a chapter *versus*
Mivart and others; I do so hate controversy, and feel I shall
do it so badly.'

Again, on July 12, he writes: 'I feel very doubtful how far I
shall succeed in answering Mivart. It is so difficult to answer
objections to doubtful points and make the discussion readable. The worst of it is, that I cannot possibly hunt through
all my references for isolated points – it would take me
three weeks of intolerably hard work. I wish I had your
power of arguing clearly. At present I feel sick of everything, and if I could occupy my time and forget my daily
discomforts, or rather miseries, I would never publish
another word. But I shall cheer up, I dare say, soon, having
only just got over a bad attack. Farewell. God knows why
I bother you about myself.

'I can say nothing more about missing links than I have
said. I should rely much on pre-Silurian times; but then
comes Sir W. Thomson[6] like an odious spectre. Farewell.'

I give these extracts because they serve to explain why
Darwin did not publish the systematic series of volumes
dealing with the whole of the subjects treated in the 'Origin'. With his almost constant and most depressing ill-
health, the real wonder is that he did so much. We can,
therefore, fully understand why, when he had published
the 'Descent of Man', in 1871, and the second editions of
that work and of the 'Animals and Plants', in 1875, with the
intervening 'Expression of Emotions', in 1872, he should

devote himself almost entirely to the long series of observations and experiments upon living plants, which constituted his relaxation and delight, and resulted in that series of volumes which are of the greatest value and interest to all students of the marvels and mysteries of vegetable life. And when, in 1881, he published his last volume upon 'Worms', giving the result of observations and experiments carried on for forty-four years, he enjoyed the great satisfaction of its being a wonderful success, while it was received by the reviewers with unanimous praise and applause.

During this latter period of his life I had but little correspondence with him, as I had no knowledge whatever of the subjects he was then working on. But he still continued to write to me occasionally, either referring kindly to my own work or sending me facts or suggestions which he thought would be of interest to me. I will here give only some extracts from a few of the latest of the letters I received from him.

On November 3, 1880, he wrote me the following very kind letter upon my 'Island Life', on which I had asked for his criticism:

'I have now read your book, and it has interested me deeply. It is quite excellent, and seems to me the best book which you have ever published; but this may be merely because I have read it last. As I went on I made a few notes, chiefly where I differed slightly from you; but God knows whether they are worth your reading. You will be disappointed with many of them; but it will show that I had the will, though I did not know the way to do what you wanted.

'I have said nothing on the infinitely many passages and views, which I admired and which were new to me. My notes are badly expressed, but I thought that you would excuse my taking any pains with my style. I wish my confounded handwriting was better. I had a note the other day from Hooker, and I can see that he is much pleased with the dedication.'

With this came seven foolscap pages of notes, many giving facts from his extensive reading which I had not seen.

There were also a good many doubts and suggestions on the very difficult questions in the discussion of the causes of the glacial epochs. Chapter xxiii, discussing the Arctic element in south temperate floras, was the part he most objected to, saying, 'This is rather too speculative for my old noddle. I must think that you overrate the importance of new surfaces on mountains and dispersal from mountain to mountain. I still believe in Alpine plants having lived on the lowlands and in the southern tropical regions having been cooled during glacial periods, and thus only can I understand character of floras on the isolated African mountains. It appears to me that you are not justified in arguing from dispersal to oceanic islands to mountains. Not only in latter cases currents of sea are absent, but what is there to make birds fly direct from one Alpine summit to another? There is left only storms of wind, and if it is probable or possible that seeds may thus be carried for great distances, I do not believe that there is at present any *evidence* of their being thus carried more than a few miles.'

This is the most connected piece of criticism in the notes, and I therefore give it verbatim. My general reply is printed in 'More Letters' iii. p. 22. Of course I carefully considered all Darwin's suggestions and facts in later editions of my book, and made use of several of them. The last, as above quoted, I shall refer to again when considering the few important matters as to which I arrived at different conclusions from Darwin. But I will first give another letter, two months later, in which he recurs to the same subject.

Down, January 2, 1881.

MY DEAR WALLACE,

The case which you give is a very striking one, and I had overlooked it in *Nature*;[7] but I remain as great a heretic as ever. Any supposition seems to me more probable than that the seeds of plants should have been blown from the mountains of Abyssinia, or other central mountains of Africa, to the mountains of Madagascar. It seems to me almost infinitely more probable that Madagascar extended far to the south during the glacial period and that the S. hemisphere was, according to Croll,[8] then more temperate; and that the whole of Africa was then peopled with some temperate forms, which crossed chiefly by agency of birds and sea-currents, and some few by the wind, from the shores of Africa to Madagascar, subsequently ascending to the mountains.

How lamentable it is that two men should take such widely different views, with the same facts before them; but this seems to be almost regularly our case, and much do I regret it. I am fairly well, but always feel half dead with fatigue. I heard but an indifferent account of your health some time ago, but trust that you are now somewhat stronger.

Believe me, my dear Wallace,
Yours very sincerely,
CH. DARWIN.

It is quite really pathetic how much he felt difference of opinion from his friends. I, of course, should have liked to have been able to convert him to my views, but I did not feel it so much as he seemed to do. In letters to Sir Joseph Hooker (in February and August, 1881), he again states his view as against mine very strongly ('More Letters', iii. pp. 25 and 27); and this, so far as I know, is the last reference he made to the subject. The last letter I received from him was entirely on literary and political subjects, and, as usual, very kind and friendly. As it makes no reference to our controversies, and touches on questions never introduced before in our correspondence, I think it will be interesting to give it entire.

Down, July 12, 1881.

MY DEAR WALLACE,

I have been heartily glad to get your note and hear some news of you. I will certainly order 'Progress and Poverty' for the subject is a most interesting one. But I read many years ago some books on political economy, and they produced a disastrous effect on my mind, viz., utterly to distrust my own judgement on the subject, and to doubt much everyone else's judgement! So I feel pretty sure that Mr George's book will only make my mind worse confounded than it is at present. I also have just finished a book which has interested me greatly, but whether it would interest anyone else I know not. It is the 'Creed of Science', by W. Graham, A. M.[9] Who or what he is I know not, but he discusses many great subjects, such as the existence of God, immortality, the moral sense, the progress of society, etc. I think some of his propositions rest on very uncertain foundations, and I could get no clear idea of his notions about God. Notwithstanding this and other blemishes, the book has interested me *extremely*. Perhaps I have been to some extent deluded, as he manifestly ranks too high what I have done.

I am delighted to hear that you spend so much time out-of-doors and in your garden. From Newman's old book (I forget title) about the country near Godalming, it must be charming.

We have just returned home after spending five weeks on Ullswater. The scenery is quite charming, but I cannot walk, and everything tries me, even seeing scenery, talking with anyone, or reading much. What I shall do with my few remaining years of life I can hardly tell. I have everything to make me happy and contented, but life has become very wearisome to me. I heard lately from Miss Buckley[10] in relation to Lyell's Life, and she mentioned that you were thinking of Switzerland, which I should think and hope that you would enjoy much.

I see that you are going to write on the most difficult political question, the land. Something ought to be done, but what, is the rub. I hope that you will (not) turn renegade to natural history; but I suppose that politics are very tempting.

With all good wishes for yourself and family,
Believe me, my dear Wallace,
Yours very sincerely,
CHARLES DARWIN.

This letter is, to me, perhaps the most interesting I ever received from Darwin, since it shows that it was only the engrossing interests of his scientific and literary work, performed under the drawback of almost constant ill-health, that prevented him from taking a more active part in the discussion of those social and political questions that so deeply affect the lives and happiness of the great bulk of the people. It is a great satisfaction that his last letter to me, written within nine months of his death, and terminating a correspondence which had extended over a quarter of a century, should be so cordial, so sympathetic, and broad-minded.

In 1870 he had written to me, 'I hope it is a satisfaction to you to reflect – and very few things in my life have been more satisfactory to me – that we have never felt any jealousy towards each other, though in some sense rivals. I believe I can say this of myself with truth, and I am absolutely sure that it is true of you.' The above long letter will show that this friendly feeling was retained by him to the last, and to have thus inspired and retained it, notwithstanding our many differences of opinion, I feel to be one of the greatest honours of my life. I have myself given an estimate of Darwin's work in my 'Debt of Science to Darwin' published in my 'Natural Selection and Tropical Nature' in 1891. But I cannot here refrain from quoting a passage from Huxley's striking obituary notice in *Nature*, summing up his work in a single short paragraph: 'None have fought better, and none have been more fortunate than Charles Darwin. He found a great truth, trodden underfoot, reviled by bigots, and ridiculed by all the world; he lived long enough to see it, chiefly by his own efforts, irrefragably established in science, inseparably incorporated with the common thoughts of men, and only hated and feared by those who would revile but dare not. What shall a man desire more than this?'

NOTES

Alfred Russel Wallace (1823–1913) made his distinctive mark on the history of British science even before discovering, in mid-century and independently of Darwin, a way to define the principle of natural selection that would answer many of the intellectual problems besetting evolutionary theory. An early misfortune caused him to lose all the specimens he had laboriously collected in the Amazon, South America (1848–52), when the ship on which he was sailing back to England sank. Undeterred, even though he had planned to sell the specimens to recover the expenses of his expedition, Wallace earned his early reputation by publishing important papers summarizing what he had learned in the Malay Archipelago. To this day the term 'Wallace's Line' is the term used to distinguish between areas: 'Oriental', e.g. Borneo and Bali (best defined in terms of the kinds of fauna to be found there) and 'Eastern' or 'Australasian', e.g. Celebes and Lombox. His lucid formulation, in 1855, of the law regulating the introduction of new species – 'Every species has come into existence coincident both in time and space with a pre-existing closely allied species' – was followed closely thereafter, in 1858, by his brilliant use of Malthus's *Essay on Population* to buttress his theory of survival of the fittest. From Ternate in the Moluccas, where he was suffering from attacks of intermittent fever, he wrote to Darwin. Until that point Wallace had been unaware of the full dimensions of Darwin's work, or the fact that Darwin had been amassing data for years in order to develop a theory along remarkably similar lines. Acting generously and high-mindedly, and with the enthusiastic concurrence of Sir Charles Lyell and Sir Joseph Hooker, Darwin arranged for the reading of Wallace's paper, together with an abstract of his own views, at the Linnean Society (1 July 1858).

When Wallace returned to England, his appreciation of Darwin's assistance in establishing his reputation led to the cultivation of a lifelong friendship. Wallace did not completely accept Darwin's interpretations of the data in his own investigations; he remained stubbornly independent of Darwin, as evidenced by the publication of perhaps his greatest work,

The Malay Archipelago (1869), and other significant publications: *Contributions to the Theory of Natural Selection* (1871), *Geographical Distribution of Animals* (1876), and *Darwinism* (1889). The full range of his interests can only be suggested. He thought seriously about spiritualism, land nationalization, and registration statistics, and published books on each of these subjects. It was altogether appropriate that, after earning the Royal Society medal in 1868, he should be awarded the Society's first Darwin medal in 1890.

1.   Henry Walter Bates (1825–92), traveller and naturalist, held the post of assistant secretary to the Geographical Society from 1864 to 1892. Darwin was very impressed by Bates's *Naturalist on the River Amazons* (1863), which set forth a sophisticated analysis of mimicry. He himself had avoided the subject in *The Origin of Species* because he was uncertain of the validity of his data; but in addition to the successful treatment of this important subject, Darwin considered it the best work on natural travels ever published in England, and wrote one of the two reviews, on this particular book, that he ever published.

2.   John Jenner Weir (1822–94), naturalist and accountant, was the Controller General of HM Customs. Darwin appreciated Weir's letters dealing with caterpillars, and described Weir as 'a very good man'.

3.   Arthur Gray Butler (1831–1909) is perhaps best remembered for managing the growth of Haileybury College, founded in 1805 by the East India Company to train its civil servants. But he was also a strong supporter of, and contributor to, literary studies, and became Dean and Tutor at Oriel College, and was select preacher before Oxford University (from 1885) as well as Whitehall preacher.

4.   Alfred William Bennett (1833–1902) specialized in alpine plants, cryptogams and microscopical research.

5.   St John Jackson Mivart (1827–1900), barrister and geologist, fiercely opposed Darwin's work. In addition to the quoted remarks about Mivart's *The Genesis of Species* (1871), Darwin finally wrote, 'I conclude with sorrow that though he means to be honourable, he is so bigoted that he cannot act fairly.'

Mivart continued his vendetta with three later books. He was excommunicated in 1900, the year he died.

6. Sir William Thomson, Baron Kelvin (1824–1907), was a physicist with strong interests in astronomy. He held the Professorship of Natural Philosophy at Glasgow University (1846–99). Darwin, who believed that the age of dinosaurs was at least 300 million years in the past, was shocked by Thomson's estimate that only 100 million years had elapsed.

7. *Nature*, 9 December 1880. The substance of this article by Mr Baker, of Kew, is given in 'More Letters', iii p. 25, in a footnote. [Wallace's footnote.]

8. James Croll (1821–90), physical geologist, became keeper of maps and correspondent for the Geological Survey of Scotland. His most influential (and controversial) book was *Climate and Time* (1875). Despite ill health, he published more than ninety papers as well as two additional books: *The Philosophy of Theism* (1857) and *The Philosophic Basis of Evolution* (1890).

9. William Graham (1839–1911) was Professor of Jurisprudence at Queen's College, Belfast. His *Creed of Science* (1881) earned Darwin's praise (in a letter to Romanes), even though Graham downplayed natural selection as the engine of social progress. Graham visited Darwin at Downe.

10. Arabella Burton Buckley (1840–1929) was Sir Charles Lyell's secretary, and visited Darwin at Down House. Darwin enjoyed her book *A Short History of Natural Science* (1876). She was on the list of personal friends invited to Darwin's funeral.

'Adam Sedgwick's Reaction to *The Origin of Species*', in *The Life and Letters of the Reverend Adam Sedgwick [II]*, edited by John Willis Clark and Thomas McKenny Hughes (Cambridge: Cambridge University Press, 1890), vol. II, pp. 356–62, 410–12 [1859–65]

In November of this year Darwin published his essay *On the Origin of Species*. One of the first copies was sent to Sedgwick with the following letter:

DOWN, BROMLEY, KENT,
November 11th, 1859.

My dear Professor Sedgwick,

I have told Murray to send you a copy of my book *On the Origin of Species*, which is as yet only an abstract. As the conclusion at which I have arrived after an amount of work which is not apparent in this condensed sketch, is so diametrically opposed to that which you have often advocated with much force, you might think that I send my volume to you out of a spirit of bravado and with a want of respect, but I assure you that I am actuated by quite opposite feelings. Pray believe me, my honoured friend,

Your sincerely obliged,
CHARLES DARWIN.

To Charles Darwin, Esq.

CAMBRIDGE, *December* 24*th*, 1859.

My dear Darwin,

I write to thank you for your work *On the Origin of Species....* [1]

If I did not think you a good-tempered, and truth-loving man, I should not tell you that (spite of the great knowledge, store of facts, capital views of the correlation of the various parts of organic nature, admirable hints about the diffusions, through wide regions, of nearly related organic beings, &c., &c.) I have read your book with more pain than pleasure. Parts of it I admired greatly, parts I laughed at till my sides were almost sore; other parts I read with absolute sorrow, because I think them utterly false and grievously mischievous. You have *deserted* – after a start in that tram-road of all solid physical truth – the true method of induction, and started off in machinery as wild, I think, as Bishop Wilkins's locomotive that was to sail with us to the moon.[2] Many of your wide conclusions are based upon assumptions which can neither be proved nor disproved. Why then express them in the language and arrangements of philosophical induction? As to your grand principle – *natural selection* – what is it but a secondary consequence of supposed, or known, primary facts? Development is a better word, because more close to the cause of the fact. For you do not deny causation. I call (in the abstract) causation the will of God; and I can prove that He acts for the good of His creatures. He also acts by laws which we can study and comprehend. Acting by law, and under what is called final cause, comprehends, I think, your whole principle. You write of 'natural selection' as if it were done consciously by the selecting agent. 'Tis but a consequence of the presupposed development, and the subsequent battle for life. This view of nature you have stated admirably, though admitted by all naturalists and denied by no one of common sense. We all admit development as a fact of history; but how came it about? Here, in language, and

still more in logic, we are point-blank at issue. There is a
moral or metaphysical part of nature as well as a physical.
A man who denies this is deep in the mire of folly. 'Tis the
crown and glory of organic science that it *does*, through
final cause, link material to moral; and yet *does not* allow us
to mingle them in our first conception of laws, and our
classification of such laws, whether we consider one side
of nature or the other. You have ignored this link; and, if
I do not mistake your meaning, you have done your best
in one or two pregnant cases to break it. Were it possible
(which, thank God, it is not) to break it, humanity, in my
mind, would suffer a damage that might brutalize it, and
sink the human race into a lower grade of degradation
than any into which it has fallen since its written records
tell us of its history. Take the case of the bee-cells. If your
development produced the successive modification of the
bee and its cells (which no mortal can prove), final cause
would stand good as the directing cause under which the
successive generations acted and gradually improved.
Passages in your book, like that to which I have alluded
(and there are others almost as bad), greatly shocked my
moral taste. I think, in speculating on organic descent,
you *over*-state the evidence of geology; and that you
under-state it while you are talking of the broken links of
your natural pedigree: but my paper is nearly done, and I
must go to my lecture-room. Lastly, then, I greatly dislike
the concluding chapter – not as a summary, for in that
light it appears good – but I dislike it from the tone of tri-
umphant confidence in which you appeal to the rising
generation (in a tone I condemned in the author of the
Vestiges)[3] and prophesy of things not yet in the womb of
time, nor (if we are to trust the accumulated experience of
human sense and the inferences of its logic) ever likely to
be found anywhere but in the fertile womb of man's imagination.

And now to say a word about a son of a monkey and an
old friend of yours. I am better, far better, than I was last
year. I have been lecturing three days a week (formerly

I gave six a week) without much fatigue, but I find, by the loss of activity and memory, and of all productive powers, that my bodily frame is sinking slowly towards the earth. But I have visions of the future. They are as much a part of myself as my stomach and my heart, and these visions are to have their antitype in solid fruition of what is best and greatest. But on one condition only – that I humbly accept God's revelation of Himself both in His works and in His word, and do my best to act in conformity with that knowledge which He only can give me, and He only can sustain me in doing. If you and I do all this, we shall meet in heaven.

I have written in a hurry, and in a spirit of brotherly love. Therefore forgive any sentence you happen to dislike; and believe me, spite of our disagreement on some points of the deepest moral interest, your true-hearted old friend,

A. SEDGWICK.

ILKLY WELLS HOUSE,
OTLEY, YORKSHIRE,
26 *November*, 1859.

My dear Professor Sedgwick,

I did not at all expect that you would have written to me. You could not possibly have paid me a more honourable compliment than in expressing freely your strong disapprobation of my book. I fully expected it. I can only say that I have worked like a slave on the subject for above twenty years, and am not conscious that bad motives have influenced the conclusions at which I have arrived. I grieve to have shocked a man whom I sincerely honour. But I do not think you would wish anyone to conceal the results at which he has arrived after he has worked, according to the best ability which may be in him. I do not think my book will be mischievous; for there are so many workers that, if I be wrong I shall soon be annihilated; and surely you will agree that truth can be known only by rising victorious from every attack.

I daresay I may have written too confidently from feeling so confident of the truth of my main doctrine. I have made already a few converts of good and tried naturalists, and oddly enough two of them compliment me on my cautious mode of expression! this will make you laugh....

I have tried to be honest in giving all the many and grave difficulties which occurred to me, or I met in published works. I cannot think a false theory would explain so many classes of facts, as the theory seems to me to do. But *magna est veritas,* and, thank God, *praevalebit.* Forgive me for scribbling at such length, and let me say again how grieved I am to have encountered your severe disapprobation and ridicule. Your kind and noble heart shews itself throughout your letter. I thank you for writing, and remain, with sincere respect,

Your truly obliged,
CHARLES DARWIN.

To Miss Gerard.

NORWICH, *January 2nd,* 1860.

...I have read Darwin's book. It is clever, and calmly written; and therefore, the more mischievous, if its principles be false; and I believe them *utterly false.* It is the system of the author of the *Vestiges* stripped of his ignorant absurdities. It repudiates all reasoning from final causes; and seems to shut the door upon any view (however feeble) of the God of Nature as manifested in His works. From first to last it is a dish of rank materialism cleverly cooked and served up. As a system of philosophy it is not like the Tower of Babel, so daring in its high aim as to seek a shelter against God's anger; but it is like a pyramid poised on its apex. It is a system embracing all living nature, vegetable and animal; yet contradicting – point blank – the vast treasury of facts that the Author of Nature has, during the past two or three thousand years, revealed to our senses. And why is this done? For no other solid reason, I am sure, except to make us independent of a Creator....

To Professor Owen.
CAMBRIDGE, Wednesday Morning [28 *March*, 1860].
My dear Owen,
...I want to pick your brains about 101 things. About Darwin's theory, about Agassiz, about the Reptiles in our (so-called) coprolite bed. By the way, I will send you a copy of last week's *Spectator*. Near the end of it is a long letter first sent to the Archbishop of Dublin[4], assuredly without any intention of its publication. But his Grace took on himself the office of man-midwife, and delivered the said brainchild to the office of *The Spectator*. The first publication fell into the hands of the *unknown Author*; who sent some corrections etc., though the staple of the letter still remained, word for word, as first written by him. For example; in one sentence the name of *Owen* appeared in the place of *Oken*, and *Pachyderms* were called *Pachydemics*[5]! I want to learn your views about creation's law. It is clear that there has been a law governing the succession of forms. But here, by *law*, I mean order of succession, and not a law like that of gravitation, out of which the actual movements of our system follow by mechanical succession. In that sense I do not believe in any law of creation. The highest point we can, I think, ever reach is a law of succession of forms, each implying a harmonious reference to an archetype, and each having indications of the action of a final cause – i.e. of intelligent causation, or creation. My belief is: 1st, that Darwin has deserted utterly the inductive track – the narrow but sure track of physical truth, – and taken the broad way of hypothesis, which has led him (spite of his great knowledge) into great delusion; and made him the *advocate*, instead of the *historian* – the teacher of error instead of the

apostle of truth: 2nd, I think that (whether he *intends* it or not, or *knows* it or not) he is a teacher of that which savours of rankest materialism, and of an utter rejection of the highest moral evidence, and the highest moral truth. I must stop for want of room,

Ever yours,
A. SEDGWICK.

Sedgwick's letter in *The Spectator* was followed up at no distant date by a direct public attack on *The Origin of Species*. He made a communication to the Cambridge Philosophical Society (7 May): *On the succession of organic forms during long geological periods: and on certain theories which profess to account for the origin of new species*. A careful synopsis of the paper, or lecture, supplied by himself to *The Cambridge Chronicle*, shews that the theory was attacked wholly from the geological side, and declared to be a mere hypothesis, at variance with the true inductive methods of attaining truth. There was a full meeting of the Society, and a lively discussion on the subject of the lecture ensued, in which several leading members took part. Henslow, though not a thorough-going partisan, defended Darwin; and Professor Clark, though disposed to agree in the main with Sedgwick, did his best to impart a philosophical tone to the discussion by suggesting that the theory ought to be classed among those imperfect inductions which point the way to truth. But the general sense of the meeting was unquestionably, we have heard, on Sedgwick's side.

To Dr Livingstone.[6]

CAMBRIDGE, *March* 16*th*, 1865.

My dear Dr Livingstone,

I have seen in one of our Papers, that you are now in London, but soon about to leave it, on a new Christian mission to Africa. Whenever, and whithersoever, you may go, may God bless your labours of love, and give you long life, and help you to perfect your benevolent plans.... I do long to have an hour's talk with you about our beloved and lamented Christian friend Bishop Mackenzie,[7] and about some other points connected with the, apparently, abandoned Mission – not, however, I trust, permanently abandoned.

I greatly dislike the tendency to formal superstitious observances in the present day. Of course I am alluding to the High Church party in England. The idolatrous element is rife amongst us. We want to lean upon our own works and merits, and count them up as if they gave us the right to draw upon our Redeemer's treasures. We are the slaves of our senses, and too willing to follow their lead, rather than to lay the foundations of holy truth in the simple teaching of the Gospel, and the acceptance of an enlightened conscience. Nor is this all. Many of our ministers and people are in an unhealthy craving for the office and power of a sacrificing Priest, and the dicta of an infallible authority, that may save them (in these days of multitudinous difficulties) from all further trouble. This might flatter the pride of the shepherd, and save both shepherd and flock from the toil of thinking, and the fatigue of further wandering. But it is false to the cause of truth, and a flinching from one of the many forms of probation that our God and Redeemer has given us. These subjects also I should perhaps have talked over with you.

Speaking too of my own craft (but age and infirmities have almost taken me out of the fields of geology) I admire the zeal with which its work is carried on; and especially the great palaeontological treasures which are spread before our admiring senses. But the Geological Society is partly in fetters. It is not the honest independent body it once was; and some of its leading men are led by the nose in the train of an hypothesis – I mean the development of all organic life from a simple material element by natural specific transmutation, ending in the flora and fauna of the actual world with man at its head. Darwin has made this theory popular, but he has not added one single fact that helps it forward; and I think that it appeared (about sixty-five years since) far better in the poetry of the grandfather, than now in the prose of the grandson. Lyell has swallowed the whole theory, at which I am not surprised – for without it, the elements of geology, as he expounded them, were illogical. He is an excellent and thoughtful writer, but not, I think, a great field-observer, and during his long geological labours he has never been able to look steadily in the face of nature except through the spectacles of an hypothesis. His mind is essentially deductive, and not inductive. Now geologists have not yet sufficiently unrolled the records of the earth to reach a starting point of knowledge from which to reason deductively with perfect safety. They may varnish it as they will; but the transmutation theory ends (with nine out of ten) in rank materialism; which is as pestilent in the investigations of material science, as is Popery in the discussions of religious truths, and the duties of a religious life. There is a world of mind, as well as a world of matter; and all the materialists on the earth will never bridge over the interval between the two. I fear I must have fatigued you, my dear and honoured friend. Often from laziness, and sometimes on principle, I abstain from letter-writing; but when I begin I never know how to stop. May God bless you, preserve you, and comfort you. So prays your aged and affectionate friend,

ADAM SEDGWICK.

NOTES

For a note on Adam Sedgwick, see above, pp. 57–8.

1. [The omitted passages contain Sedgwick's reasons for not having acknowledged the gift sooner. The whole letter is printed in Darwin's *Life*, ii. 247 – J. W. C. and T. McK. H]

2. George Wilkins (1785–1865), Archdeacon of Nottingham, was well known for his energetic good works, and much admired for his pulpit oratory.

3. A reference to *Vestiges of the Natural History of Creation* (two volumes, 1843–6), by Robert Chambers (1802–71). The work, which Chambers anticipated (correctly) would upset many of the orthodox, was at first published anonymously. Chambers was also concerned that its publication might affect adversely the financial fortunes of his publishing firm, W. & R. Chambers.

4. Richard Whately (1787–1863) was consecrated Archbishop of Dublin in 1831.

5. Sedgwick's letter appears in *The Spectator*, 24 March 1860, p. 285, introduced by the following sentence: 'The Archbishop of Dublin has received the following remarks in answer to an inquiry he had made of a friend (eminent in the world of science) on the subject of Darwin's theory of the origin of species.' It is reprinted, with considerable additions, 'revised and corrected by the author' in the same journal, 7 April 1860, p. 234. Darwin at once recognized that the article was by Sedgwick, and wrote: 'I now feel certain that Sedgwick is the author of the article in *The Spectator*. No one else could use such abusive terms. And what a misrepresentation of my notions! Any ignoramus would suppose that I had *first* broached the doctrine that the breaks between successive formations marked long intervals of time. It is very unfair. But poor dear old Sedgwick seems rabid on the question,' *Life of Darwin*, ii. 297. [Note supplied by John Willis Clark and Thomas McKenna Hughes, editors of *The Life and Letters of the Reverend Adam Sedgwick*.]

6. David Livingstone (1813–73), Scottish missionary and explorer, after a brief stay in England (July 1864 to August 1865), was raising funds to finance his return (as British Consul) to Africa. His final expedition would be considered by many to be his most important one.

7. Bishop Charles Frederick Mackenzie (1825–62) was active in Central Africa, along with Livingstone, in exploring hitherto unknown terrain and in stamping out as much of the slave trade as was feasible for the time (he personally went to war against native slavers).

'The Quarrel between Owen and Darwin', in *The Life of Richard Owen*, edited by the Reverend Richard Owen (London: John Murray, 1894), vol. II, pp. 38–40, 89–93 [1859–82]

Professor Owen's evidence before 'Mr Gregory's Committee' occupies some thirty pages in the Blue Book.[1] In it he disposes of the suggestion as to risk during removal by pointing out that two such removals had been made, under his care, at the College of Surgeons. The collections of the British Museum were, he said, mainly dried, and, therefore, would run considerably less risk in transit than the innumerable delicate preparations preserved in the collections of the College of Surgeons.

In the course of his evidence Owen made some interesting remarks concerning Darwin's work on the 'Origin of Species', just published, which helps to strengthen the impression that he was at first much taken with the new views, and felt the same friendliness toward them as he had previously shown to the views expressed in the 'Vestiges of Creation'.[2] Speaking as to the desirability of exhibiting every species, or only a proportion of the species of a group in the proposed new museum, Owen said before the committee: 'We are obliged not to have a Procrustean Law for all classes, but to be guided, as to the proportion of each class, according to the nature and significance of the differences that exist. With regard to birds, I must say that not only would I exhibit every species, but I see clearly, in the present phase of natural history philosophy, that we shall be compelled to exhibit varieties also. The whole intellectual world this year has been excited by a book on the origin of species; and what is the consequence? Visitors come to the British Museum, and they say, 'Let us see all these varieties of pigeons: where is the tumbler, where is the pouter?' and I am obliged with shame to say, 'I can show you none of them;'[3] and yet there we give what, we consider, some may think an extravagant space to the pigeons; but they are the pigeons of the whole world. As to showing you the varieties of those species, or any of those phenomena that would aid one in getting at that mystery of mysteries, the origin of species, our space does not permit; but surely there ought to be space somewhere, and, if not in the British Museum, where is it to be obtained?' The chairman of the committee said to Owen: 'I presume that the persons who make these inquiries are, to a certain extent, scientific persons?' to which he replied: 'I must say that the number of intellectual individuals interested in the great question which is mooted in Mr. Darwin's book is far beyond the small class expressly concerned in scientific research.' ...

• • •

In the latter part of 1859 Charles Darwin published his 'Origin of Species', and we gather the value he set upon Owen's opinion from the following note written to Lyell, which is included by Francis Darwin in his 'Life' of his father:

'How curious I shall be to know what line Owen will take! Dead against us, I fear; but he wrote me a most liberal note on the reception of my book, and said he was quite prepared to answer fairly, and without prejudice, my line of argument.'

After a meeting with Owen, Darwin writes him the following interesting letter respecting the 'Origin'.

Down, Bromley, Kent: December 13 (1859).

Dear Owen –... You made a remark in our conversation something to the effect that my book could not probably be true as it attempted to explain so much. I can only answer that this might be objected to any view embracing two or three classes of facts. Yet I assure you that its truth has often and often weighed heavily on me; and I have thought that perhaps my book might be a case like Macleay's quinary system.[4] So strongly did I feel this that I resolved to gave it all up, as far as I could, if I did not convince at least two or three competent judges. You smiled at me for sticking myself up as a martyr; but I assure you if you had heard the unmerciful and, I think, unjust things said of my book and to me in a letter by an old and very distinguished friend you would not wonder at me being sensitive, perhaps ridiculously sensitive. Forgive these remarks. I should be a dolt not to value your scientific opinion very highly. If my views are *in the main* correct, whatever value they may possess in pushing on science will now depend very little on me, but on the verdict pronounced by men eminent in science.

Believe me,
Yours very truly,
C. Darwin.

In the early part of this letter Darwin says he is not able to hunt up some information for which Owen has asked, as his 'notes for the latter chapters are a chaos'. The 'old and very distinguished friend' Dr Francis Darwin considers to be Adam Sedgwick.

If not 'dead against' the theory of Natural Selection, Owen at first looked askance at it, preferring the idea of the great scheme of Nature which he had himself advanced. He was of opinion that the operation of external influences and the resulting 'contest of existence' lead to certain species becoming extinct. Thus it came about, he supposed, that, like the dodo in recent times, the dinornis and other gigantic birds had disappeared. But he never, so far as can be ascertained, expressed a definite opinion on Darwinism, and in the 'Historical Sketch' which prefaces the sixth edition (1882) of 'The Origin of Species', Darwin traces Owen's ideas so far as he could comprehend them. The singular impartiality of Darwin and his increasing endeavours to arrive at the truth, whether it turned against or supported him, permit the quotation of his own words in explanation of the question.

Darwin writes: 'When the first edition of this work[5] was published, I was so completely deceived, as were many others, by such expressions as "the continuous operation of creative power", that I included Professor Owen with other palaeontologists as being firmly convinced of the immutability of species; but it appears[6] that this was on my part a preposterous error. In the last edition of this work[7] I inferred, and the inference still seems to me perfectly just, from a passage beginning with the words 'no doubt the type-form', &c.,[8] that Professor Owen admitted that natural selection may have done something in the formation of new species; but this, it appears,[9] is inaccurate and without evidence. I also gave some extracts from a correspondence between Professor Owen and the editor of the "London Review", from which it appeared manifest to the editor, as well as to myself, that Professor Owen claimed to have promulgated the theory of natural selection before I had done so; and I expressed my surprise and satisfaction at this announcement; but as far as it is possible to understand certain recently published passages,[10] I have either partially or wholly again fallen into error. It is consolatory to me that others find Professor Owen's controversial writings as difficult to understand and to reconcile with each other as I do. As far as the mere enunciation of the principle of natural selection is concerned, it is quite immaterial whether or not Professor Owen preceded me, for both of us, as shown in this historical sketch, were long ago preceded by Dr Wells and Mr Matthew.'[11]

NOTES

For a note on Sir Richard Owen, see above, p. 90.

1.   Mr Gregory, MP for Co. Galway, Ireland, headed a committee (formed in 1861), with its charge an investigation of the problems that might be created if biological specimens were to be moved from the British Museum to a new site.

2.   Robert Chambers (1802–71) published *Vestiges of the Natural History of Creation*, a pioneering work in geology, in two volumes (1843–46). Because of the controversial nature of its subject-matter, and because he did not wish to have its publication affect the financial fortunes of his publishing firm, W. & R. Chambers, he issued the work anonymously.

3.   The reader will remember that this detail of Owen's great scheme has been elaborated by Professor Sir William Flower, and is exhibited in the Central Hall of the Natural History Museum. [Richard Owen's grandson and editor of *The Life of Richard Owen*, hereafter cited as RO, wrote this footnote.] Sir William Flower, a mammologist, specialized in natural history; he was the Curator of the Hunterian Museum of the Royal College of Surgeons, and later its Director, before he became the Director of the British Museum in South Kensington. He was invited as a personal friend to Darwin's funeral.

4.   An artificial attempt at a natural system of classification which soon became a byword among naturalists. *D.N.B.* [RO]

5.   *Nature of Limbs*, 1849. Address to British Association (1858). [RO]

6.   *Anat. of Vertebrates*, vol. iii p. 176. [RO]

7.   *Origin of Species.* [RO]

8.   *Anat. of Vert.*, vol. i. p. xxxv. [RO]

9.   *Ibid.*, vol. iii. p. 798. [RO]

10.   *Anat. of Vert.*, vol. iii. p. 798. [RO]

11.   William Charles Wells (1757–1817) was born in Charleston, South Carolina, and settled in London in 1785. He published important studies of dew and binocular vision, and was an early pioneer in the field of natural selection research, applying its findings to the races of man. Patrick Matthew

(1790–1874) held pronounced political and agricultural views, as in *Naval Timber and Arboriculture* (1831) and *Emigration Fields* (1839). Darwin read the publications of both men, and paid particular attention to the Appendix of *Naval Timber*, which dealt with natural selection. He came to the conclusion, in 1865, that Wells had preceded Matthew in formulating a plausible theory of natural selection, and deserved the credit.

'A Great Friendship', in *Life and Letters of Sir Joseph Dalton Hooker, O.M., G.C.S.I., based on materials collected and arranged by Lady Hooker,* edited by Leonard Huxley, vol. I, pp. 486–503 [1839–63]

Modern Science dates from before or after the 'Origin of Species'.[1] The publication of the book was, so to say, the Hegira of Science. By it the science of living things was revolutionized and every other branch of natural science was stirred. After the vested interests of current opinion rose up in a great turmoil, Philosophy took a new element into her reckoning. The Natural Sciences claimed their rights as knowledge, discipline, and power.

But the making of the 'Origin' is not only a history of science – it is the history of a great friendship. In its fabric the two strands are indissolubly interwoven. As Darwin exclaimed to his friend, 'Talk of fame, honour, pleasure, wealth – all are dirt compared with affection, and this is a doctrine [in] which I know from your letter that you will agree from the bottom of your heart,' so the achievement is ennobled by the warm human affection that so long sustained the worker and aided the work. For twenty years the materials for the task were being amassed; for fifteen of these years Hooker was Darwin's confidant and helper. Without Hooker's aid Darwin's great work would hardly have been carried out on the botanical side.[2] In his quiet isolation at Down, cut off from the ordinary converse of the world by the perpetual uncertainties of ill-health, Darwin found refreshment and delight in pouring out to his friend his schemes of research and his wonderful experiments on the living action of plants, sure of sympathy, yet begging Hooker, if he could spare time to read these letters, at least to waste none of his too busy hours in answering them, saying:

> It is a pleasure to me to write to you, as I have no one to talk to about such matter as we write on. But I seriously beg you not to write to me, unless so inclined; for busy as you are and seeing many people, the case is very different between us (June 19, 1860). It is the greatest temptation to me to write *ad infinitum* to you (July 19, 1856).

As to direct botanical aid, he wrote with enthusiastic appreciation and careful criticism of Hooker's publications, which bore so closely on his own work. But this was the smallest part of their scientific interchange. Though he repeatedly insists 'Do not answer questions merely out of good nature' ['of which towards me you have a most abundant stock' (April 8, 1857), 'as wonderful as mesmerism' (1846)], it was the unstinted privilege of the elder friend to ask, as it was the privilege of the younger to answer from the fulness of his botanical knowledge, a host of questions bearing on the relations and distribution of individual plants and groups of plants, wherein lie answers to some of the riddles of life.

The beginnings of this friendship have been told by Hooker himself in the 'Life of Darwin', ii. 19.

> My first meeting with Mr Darwin [he tells us] was in 1839, in Trafalgar Square. I was walking with an officer who had been his shipmate for a short time in the *Beagle* seven

years before, but who had not, I believe, since met him. I was introduced; the interview was of course brief, and the memory that I carried away and still retain was that of a rather tall and rather broad-shouldered man, with a slight stoop, an agreeable and animated expression when talking, beetle brows, and a hollow but mellow voice; and that his greeting of his old acquaintance was sailor-like – that is, delightfully frank and cordial.

It has already been told how the proofs of the 'Voyage of the *Beagle*' reached him through the Lyells in the spring of that year, while he was hurrying on the last of his medical studies in order to take his degree before sailing with Ross,[3] and how, there being no other time available, he slept with them under his pillow, and read them before getting up in the morning.

> They impressed me profoundly, I might say despairingly, with the variety of acquirements, mental and physical, required in a naturalist who should follow in Darwin's footsteps, whilst they stimulated me to enthusiasm in the desire to travel and observe.

In the letters from the Antarctic there are several references to Darwin, who saw various of these letters through the Lyells. The correspondence between them, as has been told on p. 169, began in December 1843, when Darwin wrote to congratulate him on his return (C.D. ii. 21) and urged the importance of correlating the Fuegian Flora with that of the Cordillera and of Europe, at the same time offering his own collections of plants from the Galapagos Islands, from Patagonia and Fuegia for examination.

> This led to me sending him an outline of the conclusions I had formed regarding the distribution of plants in the southern regions, and the necessity of assuming the destruction of considerable areas of land to account for the relations of the flora of the so-called Antarctic Islands.

I do not suppose that any of these ideas were new to him, but they led to an animated and lengthy correspondence full of instruction.

Only the first two or three letters open with the formal 'My dear Sir' of the period; by February 1844 Darwin inaugurates 'Dear Hooker' to his 'co-circum-wanderer and fellow labourer', while from the day of his impending departure to India the 'very truly' or 'very sincerely' of either signature, gradually merging in 'Ever yours', is lost in 'Your affectionate friend' or 'Yours affectionately' maintained by both to the end.

Acquaintance ripened swiftly into friendship. 'Farewell!' Darwin concludes a letter in 1845. 'What a good thing is community of tastes! I feel as if I had known you for fifty years. Adios!" And 'forty years on' the sympathetic bond between them was as strong as ever. In 1881 Darwin writes:

Your letter has cheered me, and the world does not look a quarter so black as it did when I wrote before. Your friendly words are worth their weight in gold.

One of the starting points of Darwin's 'presumptuous work' had been the striking impression made on him by the distribution of the Galapagos organisms; hence his eager desire to know whether the botany of this isolated group was as suggestive as the zoology.

The correspondence began in December; by January the momentous confession was made:

At last gleams of light have come, and I am almost convinced (quite contrary to the opinion I started with) that species are not (it is like confessing a murder) immutable.

He had instantly recognized Hooker's capacity. 'I am pleased to think', he writes on Hooker's rejection at Edinburgh in 1845, 'that after having read a few of your letters, I never once doubted the position you will ultimately hold among European Botanists'. And in the next letter, 'It is absurdly unjust to speak of you as a mere systematist'. More than this, he recognized that Hooker also believed, as he put it in the Preface to his Flora Antarctica, that 'Geographical Distribution will be the key which will unlock the mystery of the species'.

But true views of geographical distribution were impossible without full and accurate Floras. Here no doubt was a redoubled motive for the ardour with which Hooker flung himself into his unending labours, the extent of which called forth the first of many anxious warnings from Darwin as early as 1845, to beware of overwork, doctor though he be, and a novel prescription, 'You ought to have a wife to stop your working too much, as Mrs Lyell peremptorily stops Lyell'. The perfecting of his great Floras involved the re-examination of his vast materials and the more or less incomplete work of his predecessors, so as to sweep away the existing synonymy and overlapping, and to readjust the systematic details by making clearer the true affinities and world-range of disputable genera and species. Complete and accurate classification according to nature was the first step towards finding the key to it all.

Thus Darwin, in the act of asking his aid, stimulated his native bent. He was encouraged in his inclination to deal with the wider bearings of his observations, which, in Darwin's eyes, made his Flora and his letters so different from the works of so many other systematists, remarkable for their lack of instructive general results. And though special researches such as these appeared to distract him from his main work on the Southern Floras, yet they shaped his own views and added to his reputation.

I am almost sorry for your eternal additional labours on the Galapagos Flora [writes Darwin in September 1846; but adds emphatically], as yet your work assuredly has not been thrown away, as many have referred to your curious geographical results on this archipelago.

Similarly, of a preliminary sketch of his Tasmanian results, in 1844:

I trust that your sketch will not have cost you ultimately loss of time, as, judging by myself, preliminary sketches and resketches do much good....Seriously, I almost grieved, when I saw the length of your letter, that you should have given up so much time to me. Sir William will think me a bad friend to you, but anyhow, I trust, the sketch part of your geographical results will not turn out lost time.

These generalizations gave special value to his work and led Darwin to repudiate his description of himself as not possessing a philosophic mind, 'one of the greatest falsehoods ever told by implication; read your own Galapagos paper and be ashamed of yourself' (the whole passage is given in C.D. ii. 37). In short (March 31, 1845):

Nothing would do you so much good as a little vanity, and then you would not talk of collecting facts for others, when, say just what you please, I am sure no one could put them to better use than yourself.

It was a unique relationship of minds. Each had had the same kind of experience in world-travel, and had observed nature, animate and inanimate, with a special interest in the same question – namely, how the different forms of life had reached their present habitats. In this, indeed, the younger man had taken the elder for his model. Before their friendship and alliance began, Darwin, the born scientific enquirer with philosophic breadth of mind albeit small technical training, had advanced far along his memorable line of research. He took everything for his province that bore on heredity and variation, fertility and decline in living forms, the competition they had to meet, their range and movement, the relation of them to their fossil predecessors in the same area, the geological changes which had determined the ancient courses of migration. Hooker, master of a whole branch of science, with technical training in it from his childhood up, and equally awake to the part played by geologic change in the problems of distribution he longed to solve, eagerly placed his vast knowledge, his sound criticism, his special observation during his later travels, at the disposal of the inspiring friend and fellow-worker who had gone so much further on the same quest as himself and had pushed it into wider fields than his own.

Each was deeply conscious of his debt to the other. Of the discussions they used to have, Hooker records ('Life and Letters of Charles Darwin', ii. 27): 'I at any rate always left with the feeling that I had imparted nothing and carried away more than I could stagger under.' Darwin from the earliest time feels the immense value of his help, in books lent, summaries of results, in published works, letters, conversations. 'For my own part,' he writes after a visit of Hooker's to Down, 'I learn more in these discussions than in ten times over the number of hours reading.' And again, after reading the Antarctic Flora, he speaks of having 'extracted more facts and views from you than from any other person', while 'my pen runs away with me when writing to you' (March 19, 1845).

The thanks of a later period are foreshadowed by the thanks of the first twelvemonth, as:

Really I do not know how to thank you half for all you have done for and sent to me. I might with truth do so for every single paragraph in your letter and every one volume.... Your remarks are exactly the thing, which ever since being in Tierra del Fuego, I have felt a keen curiosity about, and have often complained to Henslow how rarely I could find any such general remarks in Botanical works.

And in 1845 the prospective break in their personal intercourse, if Hooker were elected to the chair at Edinburgh,

is a heavy disappointment to me; and in a mere selfish point of view, as aiding me in my work, your loss is indeed irreparable.... I assure you deliberately that I consider all the assistance which you have given me is more than I have received from anyone else, and is beyond valuing in my eyes.

More than this: they can express themselves with animation to each other, without risk of being misunderstood.

Hooker contributes much from his own knowledge. Distribution is his favourite subject, and he supplies statistics in the form desired to show range and migration, struggle and survival, from the Floras of the Southern Hemisphere or India or the Polar regions, all of which have fallen within his direct research. Moreover, he is particularly able to tell much about variation, for, as the preceding chapters show, he had long been struck by the incertitude of botanists on this head, and comparing detailed results all over the vast fields he had covered, had found many species as defined by local observers to be but varieties of a common species with every intermediate gradation. He can put Darwin in the way of answering the question whether large genera with wide ranging species, as should be the case with strong and increasing kinds, produce more varieties than smaller groups. At the same time he adds a warning as to the different impression of distinctness made on botanists by a given degree of difference occurring within the large or small group, so that what here would be ranked as a variety, would there be ranked as a species, to the confusion of any statistics that merely compare the relative numbers in existing lists. This is one of the cases where Hooker, after raising all the possible objections which must be overcome, is himself converted to Darwin's view by the facts which he has elicited for him.

He vehemently repudiates the notion (suggested by a geological article) of coal having been formed in shallow seas, and about this Darwin long continues to poke fun at himself and the botanists, to whom he finds it is the proverbial red rag. They differ as to continental extensions. While both condemn Forbes' unrestrained speculations in this direction,[4] Hooker is too liberal for Darwin, who, though on occasion claiming and accepting great geological changes in land and sea, stands out against volcanic islands in the ocean being thus linked to continents, or the invocation of vast upheavals and depressions without other and independent evidence, as a simple way of accounting for a single phenomenon in distribution. Later, however, we see him constrained to accept Hooker's claim for a continental extension to New Zealand, as one of the cases that 'required it in an eminent degree' but through a vanished Antarctic land, not directly to Australia.

Meantime he debates with his friend every other possible form of transport. Seeds may be carried by winds, ocean currents, berg transport, in mud clinging to a bird's foot, in the crops of birds, even the most unexpected birds, as when to his triumph a petrel is found helping in the transport of certain nuts. He confounds the popular belief that seeds of every kind must inevitably be destroyed by immersion in sea-water, through a series of experiments on temperate and tropical seeds, the latter supplied often from Kew, where also some of the experiments are repeated. He makes a salt-water tank, and tests the power of seeds to sink or swim, discovers how many will germinate happily after this treatment. He tells how his children at Down anxiously watch the trials to see whether he will 'beat Dr Hooker'. Then as the experiments proceed and a seed to be experimented on happens to be delayed, he chaffs his friend merrily: 'I believe you are afraid to send me a ripe Edwardsia pod for fear I should float it from New Zealand to Chile!'[5] And so he quickly routs Hooker's cautious scepticism. The latter, confident that nothing will happen, has planted some seeds that the Gulf Stream has carried across the Atlantic to the coast of Norway. They germinate perfectly, and in answer to his confession of defeat (the letter is not extant), Darwin writes (June 1, 1856):

> I read your note as far as 'unutterable mortification' and was in despair, for I came instantly to the conclusion that probably Government had determined to give up Kew Gardens! and you may imagine how I laughed when I

came to the real cause of mortification. It is the funniest thing in the world that you do not rejoice; for you have (as I never have) put in print that you do not believe in multiple creation, and therefore you surely should rejoice at every conceivable means of dispersal. Well, I and my wife have enjoyed a jolly laugh, and all the more from fully believing for a second that some great calamity had befallen you.

To quote a few more of the points with which the letters teem: Does the evidence show that in plants as in animals variability increases in parts which are abnormally developed? Do experiments in the Kew greenhouses show that cross fertilization improves the fertility of the plant? Do statistics indicate that trees, where the abundance of adjacent blossom would tend to self-fertilization, counteract this tendency by being more often dioecious than other plants? What of hybridism in botany; or of the part played by insects in fertilization? On what definition does a botanist rank a class of plants as high or low in the scale, and how is competitive highness measured, i.e. that superiority in development which enables, say, the recent forms of Europe and Asia to oust Australian forms when they meet, especially as some particular adaptations in a 'high' class represent a retrogression according to the usual standard, which measures 'highness' by increasing complexity of structure? How far do physical conditions alone effect similar changes in different plants? How far do the curious facts of distribution among Arctic plants indicate an extended glacial climate? Does the evidence from the migration and variation of temperate and subarctic plants indicate that this cold spell, was world-wide, and was a factor in producing 'representative species' now isolated from each other?

Without further quotation of detail here is enough to illustrate the range of Hooker's abounding help in matters of fact or of theory. Unfailing also is his information about books to be consulted or papers in scientific journals dealing with special points. Many were not procurable even from the Linnean Library,[6] where Hooker arranged that Darwin could take out what volumes he wanted. Many he lent to his friend from his own botanical library to be studied and lightly marked on the margins for the purposes of his analysis, sometimes to be borrowed afresh that the marked passages might be consulted anew when some better scheme of analysis had presented itself or some flaw had been detected in the previous scheme. 'I never cease begging favours of you,' writes Darwin in August 1855, when asking for the loan of the copy he had before of Asa Gray's Manual.

The parcels generally go from Kew to the Nag's Head in the Borough, the headquarters of the Down carrier, whether botanical parcels or a 'magnificent and awful box of books', though in the case of a rare orchid in flower, Parslow, the immemorial butler, would travel to Kew and carry it back in his own safe hands.

Once, when Hooker had a fair copy of one of Darwin's MSS. to read, a misfortune happened which recalls, though it happily did not equal, the catastrophe to the sole MS. of Carlyle's 'French Revolution' in J. S. Mill's house. The bundle 'by some screaming accident' had got transferred to the drawer where Mrs Hooker kept paper for the children to draw upon – and they 'of course had a drawing fit ever since'. Nearly a quarter of the MS. had vanished when Hooker prepared to read it at the end of a busy week.

> I feel brutified, if not brutalized [he confides in Huxley that evening], for poor D is so bad that he could hardly get steam up to finish what he did. How I wish he could stamp and fume at me – instead of taking it so good-humouredly as he will.

Nor did Hooker merely leave to his friend the tabulation of these important statistics of variation and distribution from the sources thus supplied. He often undertook it himself as a side-work in the flora on which he was at work, whether of New Zealand or India or Australia or the Arctic regions, for no other worker and no published book could provide the answer.

By a happy compensation these free gifts of time and labour for friendship's sake brought their own reward. With Hooker, as with others, such as Asa Gray,[7] whose opinion Darwin had asked on similar points, the consequent research independently enriched his own books, widened the scope of his results, and pointed the way to a revivifying theory. Writing to Hooker in January 1857, Darwin says:

> You know how I work subjects, namely if I stumble on any general remark, and if I find it confirmed in any other very distinct class, then I try to find out whether it is true, if it has any bearing on my work.

From this sprang many of his special researches. It was an additional merit in his procedure that he not only saw the crucial points that needed investigation, but inspired his most open-minded friends to independent research on the same lines, leading them to generalize on their results, instead of resting content with mere statements of fact. Thus, when Hooker writes (in December 1857):

> I have begun my Introd. Essay to Tasmanian Flora. I think I shall confine it to a clear exposition of all the main features of the Flora of Australia and leave all conclusion drawing to others:
> I am very sorry [he replies] to hear you do not intend to give generalizations in your Tasmanian Introduction but I do not believe you will be able to resist; what is in the spirit must come out.

Happily this resolve was broken by the impulse of Darwin's compulsory publication.

However, Hooker's long established conviction that species are more variable and less easily defined than most naturalists believed, did not bring him at once into the Transmutationist camp. He accepted the considerable variability of species and their spread by migration each from some one original starting place, a point less difficult perhaps to define than the perplexing modes of migration: he accepted even their relationship to allied species, their fossil predecessors in the same area, but to accept so much was not to accept their transmutation from other species. He went to India 'possessed, but not converted' by Darwin's theories, and was somewhat disappointed not to find them cleared up by the discovery of transitional forms in Sikkim, the meeting ground of tropical and arctic flora. The actual process of transition had not been observed; the partial light thrown on the question in fragmentary discussions was not enough, and until 1858–9, after the consolidation of Darwin's arguments in the famous Abstract, Hooker, as has been already noticed, worked avowedly on the accepted lines of the fixity of species, for which he had so far found no convincing substitute.

His critical attitude so long maintained may be regarded less as opposition to the tendencies of Darwin's speculations, than as the caution of a judicial mind, that required wholly convincing proof for itself before accepting the theory and all its consequences, and was equally desirous that the proof be wholly convincing for the credit of the friend who advanced it. Darwin never tires of telling how he values his criticisms. They led not to destruction, but to reconstruction. 'You never make an objection without doing much good,' he exclaims (November 18, 1856). After a long talk together, 'fighting a battle with you clears my mind wonderfully' (October 19, 1856), or, touching Hooker's help over the question of large genera varying largely, already mentioned, 'Again I thank you for your valuable assistance.... Adios, you terrible worrier of poor theorists!'

But as long as the full argument of the 'Origin' had not been presented in consecutive form, there was the constant probability that criticism on a single point could not know that it was already outflanked by a previous argument, developed elsewhere by the author, but not impressed on the critic in this particular connection. Thus replying to Hooker, who finds the changes effected by external conditions inconstant and unequal to modifying species, Darwin urges (November 11, 1856) that the external conditions by themselves do very little in producing new species, except as causing mere variability upon which selection can work. He feels strongly that to make this clear, he ought to have sent Hooker a preliminary note on variation and its causes.

In this connection it may be noted that even after the publication of the 'Origin' Hooker continued to lay more stress on external conditions than did Darwin, who explains (May 29, 1860) that he sees in almost every organism (though far more clearly in animals than in plants) *adaptation*, and this, except in rare instances, must, he thinks, be due to selection.[8]

Again (March 16, 1858) Darwin finds the reason for various difficulties raised by Hooker in the fact that probably he has not yet sufficiently explained his notions, and begs his friend to await the MS. dealing with these points. So when he does send fairly complete sections of his MS. to his chief critic, his words, 'Believe me I value to the full every word of criticism from you, and the advantage which I have derived from you cannot be told,' are a measure of the delight and relief at that critic's appreciation of the finished argument. The process bears out the phrase of June 2, 1857:

> Although we are very apt, I have observed, at the first approach of a subject, to take different views, we generally come to a near approach after a talk.

Indeed, in writing on the subject, Darwin confesses, 'I try to give the strongest cases opposed to me. I have been working your books as richest (and vilest) against mine' (July 12, 1856). But in the end, when the first paper expounding his views had been read at the Linnean, he concludes:

You cannot imagine how pleased I am that the notion of Natural Selection has acted as a purgative on your bowels of immutability. Whenever Naturalists can look on species changing as certain, what a magnificent field will be open, – on all the lines of variation – on the genealogy of all living beings – on their lines of migration, &c., &c.

At the end as at the beginning he was keenly aware of all the help Hooker had lent, help which, as has been said, Hooker himself rated at nothing. Darwin, however, exclaims:

You speak of my having 'so few aids'; why should you...[you] yourself for years and years have aided me in innumerable ways, lending me books, giving me endless facts, giving me your valuable opinion and advice on all sorts of subjects, and more than all, your kindest sympathy.

Again, when the Abstract had been set going after Wallace's paper had come like a bolt from the blue,[9] he cries, 'in how many ways have you aided me.' Yet again, when this delicate situation had been arranged, he adds, 'You must let me once again tell you how deeply I feel your generous kindness and Lyell's on this occasion; but in truth it shames me that you should have lost time on a mere point of priority.' Still, perhaps the greatest service of all was 'making me make this abstract; for though I thought I had got all clear, it has clarified my brains much, by making me weigh relative importance of the several elements,' and 'I shall, when it is done, be able to finish my work with greater ease and leisure.'

Perhaps the most remarkable tribute paid by Darwin to his friend is that which is given in the 'Life and Letters', ii. 138. The date is October 1858, while he was hard at work on the Abstract. Hooker the critic had seemed strangely unmoved by the arguments advanced, but a rather despondent note praying him not to pronounce too strongly against Natural Selection till he had read the Abstract, brought an enthusiastic reply, declaring that Darwin's speculations had been a 'jampot' to him. To this Darwin rejoins:

> I wrote the sentence without reflection. But the truth is I have so accustomed myself, partly from being quizzed by my non-naturalist relations, to expect opposition and even contempt, that I forgot for the moment that you are the one living soul from whom I have constantly received sympathy. Believe that I never forget even for a minute how much assistance I have received from you.

But Darwin, with his usual generosity of spirit, watching the increasing parallelism of their views, feared lest he had checked Hooker's original thoughts by discussing his own views with him so fully and freely. Hooker would have been the last to admit anything of the sort. He, as has been said, while gradually loosening the foundations of his former opinions, was slow to reach conviction as to the new, and only under stress of the completed argument of the 'Origin'. His original interest in their common problems connected with Geographical distribution and the unsatisfactory views current about species, was ever intensified by their constant discussions, while the special investigations, the result of which often helped to push him along the Darwinian path, were frequently prompted or stimulated by Darwin's enquiries. His own ideas involved mutability of species. Yet so long as he remained unpersuaded of a true cause for mutability, he could hardly have carried these ideas to their full completion.

Darwin's feeling, well expressed in the letter of December 25, 1859, which is given in the 'Life and Letters', ii. 252, appears further from an as yet unpublished passage in his letter of November 14, 1858, the remainder of which is given in C.D. ii. 139 and M.L. i. 455.

> I have for some time thought that I have done you an ill-service, in return for the immense good which I have

reaped from you, in discussing all my notions with you; and now there is no doubt of it, as you would have arrived at the mixture [?] independently. My only comfort is, that without you were prepared to give up species, you must have been greatly bothered in your conclusions, for the ranges of identical and representative species are so mixed up in this case, as hardly to be separated. And I can most truly say that I never thought that I might be interfering with your independent work.

And again, on January 28, 1859:

I never did pick anybody's pocket, but whilst writing my present chapter [Geographical Distribution] I keep on feeling (even while differing most from you) just as if I were stealing from you, so much do I owe to your writings and conversation: so much more than mere acknowledgements show.

Hooker, however, took the opposite view in the missing letter to which Darwin replies on April 2:

Do not fear about interfering with me in your publications. I have little doubt your views will be, and have arisen, independent of mine.

[And on Ap. 7,] The Fl. Austr. and Origin contain much of the same, but yet somehow everything is taken up from such different points of view, that I do not think we shall injure the originality of our respective books.

[In short,] You may say what you like, but you will never convince me that I do not owe you *ten* times as much as you can owe me (Dec. 30, 1858, M.L. i. 117).

But Hooker would never admit this, and five years later, when Lyell, in his forthcoming 'Antiquity of Man', proceeded to give him large credit for his services to the Darwinian theory, his native impulse was to send Darwin a flat disclaimer (March 15, 1863):

He has written to me also about the date of publication of the Australian Essay, as preceding your 'Origin' – in this matter he has got into a fix by giving said Essay a prominence which in the history of the discussion it (and its author) do not deserve. I have such an extreme aversion to intrude myself personally into such matters, and such an abomination of reclamations, that I cannot set him right, even did the plan of his book now admit of his giving the Essay less prominence. As it is, I am ashamed of seeing it paraded with an italicized heading, just as you and the 'Origin' are, and an importance given to its priority of publication which it never dreamt of claiming. Had I really believed that your 'Origin' would have been out so soon after it I really think I should have delayed the Fl. Tasmanica rather than antedate you; but though I knew you were actually printing the 'Origin', I knew how long it had been delayed, I knew how uncertain your health was, and I was working myself to death to get the Tasmanian Flora and its (for me) gigantic expenses off my hands. As it is Lyell seems to think me entitled to a goodly share of the credit of *establishing*, though not *originating*.

1. Because of your over-generous acknowledgement of assistance from me in the 'Origin'.

2. Because it was my making him eat the leek of variation, that so stupefied his senses that he was enabled to swallow Origin and apply Selection (as gastric juices).

3. Because I forced the card of non-reversion of varieties.

4. Because I first applied many of your results to the class and district of one Flora and country, in a way intelligible to him.

5. Because he understood my arrangement of the subject better than yours – at least so he said, some 18 months ago.

All this is no reason for putting me *in the same category with you as propounder* of the doctrine, which his work seems to me too much to do. However, I have not alluded to this subject to him, nor should I, if he had been as careful never to mention my name, as Huxley would seem to be, not that he really is so in the least I am sure.

To this Darwin replied (March 17):

What a candid honest fellow you are, too candid and too honest. I do not believe one man in ten thousand would have thought and said what you say about your own work in your letter. I told Lyell that nothing pleased me more in his work than the conspicuous position in which he very properly placed you.

NOTES

Sir Joseph Hooker (1817–1911) was the younger son of Sir William Jackson Hooker, the Regius Professor of Botany at Glasgow University and an eminent scientist in his own right.

Joseph Hooker began his scientific career the same way Darwin did, as the designated naturalist on a ship, in this case the *Erebus*, sailing on an Antarctic expedition in 1839. (Later in the century Thomas Huxley followed a remarkably similar path.) Hooker, like Darwin, kept meticulous notes in his journal, and collected specimens zealously.

Darwin's personal relationship with Hooker began early, and lasted a lifetime. From a close reading of Hooker's letters sent back to England from the *Erebus*, to frequent personal visits over the fifteen-year period that followed, to tough-minded exchanges of views that confirmed Hooker's interest in the geographical dispersion of species as well as his own hammering-out of the principle of natural selection, Darwin enjoyed his friendship with Hooker more than with any other scientist; he said as much on more than one occasion.

Hooker's position as botanist on the Geological survey, his travels in Syria, Morocco, the United States and Eastern Asia, his prolific publications, and the Directorship of Kew certified his right to deal as an equal with Darwin's evolving theories, and to dissent vigorously from their assumptions or conclusions whenever he believed that the supporting documentation was inadequate. It was Hooker who counselled Darwin how to deal with Russel Wallace's upsetting letter of 1858, in which the possibility of Wallace's anticipating, by publication, major elements of the research and conclusions of *The Origin of Species* loomed large as a threat to Darwin's great work, as yet unfinished. Leonard Huxley's judicious treatment of the correspondence identifies why and how the two men learned so much from each other. It may well be described as one of the finest biographies published during the Victorian Age.

1.   The full title: *On the Origin of Species by Means of Natural Selection, or the Preservation of Favoured Races in the Struggle for Life* (London: Murray, 1859).

2.   Sir F. Darwin and Professor Seward, in [*More Letters*] i. p. 39. [Leonard Huxley's footnote, hereafter cited as LH.]

3.   Sir James Ross, a merchant navy captain, first met Darwin on Direction Island, which formed part of Cocos Keeling Islands (owned by Ross), in 1833. In 1842–3 Ross led an Antarctic expedition with Joseph Hooker aboard as ship's surgeon and botanist; he turned over to Darwin his Antarctic specimens. Later, Ross searched for the missing *Erebus* and *Terror*, which were lost in the Canadian ice-pack while attempting to find a navigable Northwest Passage.

4.   This casual remark is unfair to the memory of Edward Forbes (1815–54), an extraordinarily active and productive scientist. His research on the natural selection and geographical distribution of species earned him wide respect; he was also a pioneer in elucidating fundamental concepts in palaeontology.

5.   The plant is only found in these two countries. It was shown that leguminous seeds as a rule were destroyed by immersion, thus suggesting a reason for the peculiarities in the distribution of the Leguminosae. [LH]

6.   The Linnean Society of London maintained a library that Darwin (elected a Fellow in 1854) used frequently. John Collier's oil portrait of Darwin hangs in the Society rooms at Burlington House.

7.   Asa Gray (1810–88), an American botanist, met Darwin at Kew shortly before 1855 (the date is uncertain). In 1859

Darwin sent him a first edition copy of *Origin*. Darwin was so highly pleased by Gray's article 'Natural Selection not Inconsistent with Natural Theology' (published in *The Atlantic Monthly*, October 1861) that he paid for its reprinting as a pamphlet; he personally distributed thirty-two copies. In 1877 Darwin dedicated to Gray his latest publication, *Forms of Flowers*.

8.   Thus, in March 1862, Hooker wrote to Bates: 'I am sure that with you as with me, the more you think the less occasion you will see for anything but time and natural selection to effect change; and that this view is the simplest and clearest in the present state of science is one advantage, at any rate. Indeed I think that it is, in the present state of the inquiry, the legitimate position to take up; it is time enough to bother our heads with the secondary cause when there is some evidence of it or some demand for it – at present I do not see one or the other, and so feel inclined to renounce any other for the present.' Hereupon Darwin finds it 'curiously satisfactory' to see him and Bates 'believing more fully in Natural Selection than I think I even do myself'; but he startled Darwin in November with the frank confession that every single difference which we might see might have occurred without any selection, having got right round the subject and viewed it from an entirely opposite and new side. 'I do and have always fully agreed,' is Darwin's answer, but under certain provisos, which in fact do not seem to occur. See M. L. i. 212, 199, and 223. [LH]

[Henry Walter Bates (1825–92), the 'Naturalist on the Amazons'. His boyish zeal for entomology took fuller shape under the inspiration of A. R. Wallace, with whom he set out in 1848 for those unharvested regions. Here he spent eleven years. His wide researches into the insect fauna and the problems of mimicry led him towards the theory of natural selection, and he became at once a staunch supporter of Darwin when he returned in 1859. From 1864 until his death he was Assistant Secretary to the Royal Geographical Society, and he was elected FRS in 1881.] [LH]

9.   It will be remembered how Wallace, on realizing the vast work already done by Darwin to establish the theory on an incomparably broader basis than the observations which had suggested the same theory to himself, generously waived all claim to priority. When in May 1864, in his paper on the

Evolution of Man, in the *Anthropological Review*, he repeated his disclaimer, Hooker writes to Darwin (14 May): 'I am struck with his negation of all credit or share in the Natural Selection theory – which makes one think him a very high-minded man.' [LH]

'Sir Charles Lyell and *The Origin of Species*', in *Life, Letters and Journals of Sir Charles Lyell, Bart.*, edited by Mrs Lyell [Sir Charles Lyell's sister-in-law] (London: John Murray, 1881), vol. I, 43, 45, 325–6, 358, 383–4 [1838–64]

To Charles Darwin, Esq.

Kinnordy: September 6, 1838.

My dear Darwin, – I must first read your letter again which I answered in a great hurry at Newcastle. I should like to have a talk over Salisbury Craigs with you, especially on the spot. I do hope some day that we shall be able to examine together some of the volcanic rocks on the coast near the Red Head, within a day's ride of this place. It is a splendid exhibition, and I think we should make out several points of eruption and sections of the feeders of the old volcanic islands of the Old Red Sandstone period. The variety of porphyries and amygdaloids is quite splendid. I assure you my father is quite enthusiastic about your journal, which he is reading, and he agrees with me that it would have had a great sale if separately published. The other day he told me that he wished to get a copy bound the moment it was out, and send it as a present to Sir William Hooker,[1] who more than any one would be delighted with yours. He was disappointed at hearing that it was to be fettered by the other volumes, for although he should equally buy it, he feared so many of the public would be checked from doing so. I hope you mean to sell a portion of those copies which I think you told me you were to have separate, as I think it was a large number. . . .

My father[2] has been more and more taken up and delighted with your journal, and begged me this morning to invite you to come here any day this or the next month, when we shall be here, for as long or short a time as you like. Steam-boats every Wednesday to Dundee, passage from thirty-six to forty hours; railroad four times a day from Dundee to Glamis, where the carriage meets you, and brings you in half an hour to Kinnordy – an easy trip for one who was never sea-sick except in sailing vessels. Do come, if you want some fresh air, and if you choose to bring MS. here and write, as I do whenever I choose, four or five hours quietly every day, I promise you the means of doing so; or if you prefer a geological excursion, remember that an autumn on this East coast may be almost always reckoned upon for fine weather. . . .

To Charles Darwin, Esq.

Drumkilbo, Meigle, Perthshire: October 3, 1859.

My dear Darwin, – I have just finished your volume [Origin of Species], and right glad I am that I did my best with Hooker to persuade you to publish it without waiting for a time which probably could never have arrived, though you lived to the age of a hundred, when you had prepared all your facts on which you ground so many grand generalizations.

It is a splendid case of close reasoning and long sustained argument throughout so many pages, the condensation immense, too great perhaps for the uninitiated, but an effective and important preliminary statement, which will admit, even before your detailed proofs appear, of some occasional useful exemplifications, such as your pigeons and cirripedes, of which you make such excellent use.

I mean that when, as I fully expect, a new edition is soon called for, you may here and there insert an actual case, to relieve the vast number of abstract propositions. So far as I am concerned, I am so well prepared to take your statements of facts for granted, that I do not think the *pièces justificatives* when published will make much difference, and I have long seen most clearly that if any concession is made, all that you claim in your concluding pages will follow.

It is this which has made me so long hesitate, always feeling that the case of Man and his Races and of other animals, and that of plants, is one and the same, and that if a *vera causa* be admitted for one instant, of a purely unknown and imaginary one, such as the word 'creation', all the consequences must follow.

I fear I have not time to-day, as I am just leaving this place, to indulge in a variety of comments, and to say how much I was delighted with Oceanic Islands – Rudimentary Organs – Embryology – the Genealogical Key to the Natural System – Geographical Distribution; and if I went on I should be copying the heads of all your chapters.

With my hearty congratulations to you on your great work,

Believe me ever very affectionately yours,

CHARLES LYELL.

TO CHARLES DARWIN, ESQ.

Freshwater Gate, Isle of Wight: August 20, 1862.

My dear Darwin, – Mr Jamieson of Ellon[3] has been again to Lochaber, and confirms his former theory of the glacier lakes. The chief new point is a supposed rise at the rate of a foot per mile of the shelves as we proceed from the sea inland. It seems to me to require many more measurements, before we can rely on it. He found some splendid moraines opposite the mouth of Glen Trieg. He found some shells of Arctic character in the forty feet high raised beach of the Argyllshire coast, and has asked me to learn about one of them, of which he sends a drawing.

I fell in yesterday in my walk with Mr A. G. More,[4] whom you cite in your orchid book. He considers you the most profound of reasoners, to which I made no objection, only being amused at remembering that, such being the case, you had performed a singular feat, as the Bishop of Oxford assured me, of producing 'the most illogical book ever written'.

We shall be here for a week longer. I have been with my nephew Leonard to Alum and Compton Bays.

Ever most truly yours,
CHARLES LYELL.

To CHARLES DARWIN, ESQ.

53 Harley Street, London: November 4, 1864.

My dear Darwin, – I was delighted to hear yesterday at the Athenaeum that the Council had decided that you were to have the Copley medal, for when it was not awarded to you last year I felt that its value had been much lowered, and in my indignation at the want of courage implied in their hesitation, I sympathized with a friend who has long held that these medals do more harm than good, which, however, I have always been unwilling to believe.

In the present instance it is of more than usual importance, not in a purely scientific point of view, for your reputation cannot be the least raised by it in the minds of those whose opinions you care for, or who are capable of judging for themselves as to the merits of such a book as the 'Origin', but because an honour openly conferred by an old chartered institution acts on the outsiders and helps to increase that stock of moral courage which is so small still, though it has grown sensibly in the last few years. Huxley alarmed me by telling me a few days ago that some of the older members of the Council were afraid of crowning anything so unorthodox as the 'Origin'. But if they were so, they had the good sense to draw in their horns.

Believe me ever affectionately yours,

CHARLES LYELL.

NOTES

Sir Charles Lyell (1797–1875) performed herculean labours in developing the science of geology, triumphing over weak eyes that plagued him from the beginning and finally failed him completely. His three major works – *The Principles of Geology* (1830–2), *Elements of Geology* (1838 and many revisions thereafter), and *The Antiquity of Man* (1863) – stirred controversy (among other things, they put paid to the catastrophic theories so popular among his predecessors and contemporaries), but finally became established as the most authoritative treatments of the science. *Elements of Geology* was used as the basic text for students of geology for most of the Nineteenth Century.

Darwin's interest in the differences between successive faunas relied heavily on the research recorded in Lyell's publications. Moreover, Darwin's openness to ideas that suggested the need for re-examining the platitudes of palaeontology (for instance, the idea that birds and mammals could not have coexisted in earlier time-periods because species did not evolve continuously; the evidence for their coexistence was debatable, or at least, in the Scottish phrase, not proven), were shaped directly by Lyell's interpretation of the evidence he had so assiduously – and personally – gathered. Lyell's endorsement of uniformitarianism was sometimes viewed as a refutation of Darwinism, but in fact it was aimed more directly at Lamarck's theory of transmutation of species. Lyell was so impressed by the quality and the similarities of the data gathered by Russel Wallace and Darwin that he and Joseph Hooker arranged for the joint presentation and publication of their research. His warm support of *The Origin of Species* in the months following its publication was deeply appreciated by Darwin. (Later, nevertheless, Darwin expressed reservations about Lyell's conservative interpretations of data in *The Antiquity of Man*.) Darwin recognized Lyell's importance as the foremost geologist of the century; he also admired Lyell's liberal religious ideas, and wrote of them in his *Autobiography*.

1. Sir William Jackson Hooker (1785–1865) achieved his first success when he published *Tour in Iceland* (1809). Its very publication was a *tour de force*, inasmuch as he was writing from memory; the ship to which he had entrusted his natural history specimens had burned on the homeward voyage, and he had almost died in the conflagration. By the late 1830s, when Lyell wrote the letter to Darwin quoted here, Hooker's herbarium enjoyed an international reputation. Moreover, Hooker wrote a definitive treatment of the mosses of Great Britain and Ireland, and had become a highly respected professor at Glasgow University.

2. Charles Lyell (1767–1849), translator of major works by Dante, was a famous botanist.

3. Thomas Francis Jamieson (1829–1913), of Ellon, Aberdeenshire, was a geologist who, among other accomplishments, solved the problem of the Parallel Roads of Glenroy.

4. Alexander Goodman More (1830–95), botanist, lived on the Isle of Wight, and, after serving as an assistant at the British Museum (1867–80), was appointed Curator of Botany (1881).

'A Family Friend Comments on *The Origin of Species'*, in *Harriet Martineau's Letters to Fanny Wedgwood*, edited by Elisabeth Sanders Arbuckle (Stanford, Calif.: Stanford University Press, 1983), pp. 186–93, 200–2 [1860]

To Erasmus Darwin

Ambleside
Feb^{ry} 2^d 60

Dear friend

...Well, but, what I write for is to thank you again for sending me your brother's book.[1] As for thanking *him* for the book itself, one might say 'thank you' all one's life without giving any idea of one's sense of obligation. It has been an immense pleasure to Maria and me; and, I need not add, much more than a pleasure. I am not pretending to speak about the science; though I fancy I follow his argument as a learner. If we could follow no further, the unconscious disclosure of the spirit and habits of the true scientific mind would be a most profitable and charming lesson to us. I believed, and have often described, the quality and conduct of your brother's mind; but it is an unspeakable satisfaction to see here the full manifestation of its earnestness and simplicity, its sagacity, its industry, and the patient power by which it has collected such a mass of facts, to transmute them by such sagacious treatment into such portentous knowledge. I should much like to know how large a proportion of our scientific men believe that he has found a sound road to the upper ranges of the history of organized existence. It does not very much matter; for it is the next generation that effectively profits by such works: but it would be pleasant to know that a good many remain openminded. Nobody trusts Owen's manner or speech on such occasions: but there is some wonder about what he thinks.[2] At Roebuck's[3] he spoke so as to lead hearers to believe that he sanctioned the theory of the book: whereas, on another occasion (the Geological Society?) he opposed it as untenable. But these individual cases don't matter much, when the work has fairly obtained a hearing. The review in the *Times* was by a literary man, with high scientific sanction and (I believe) assistance.[4] Who did 'D. News' I don't know:[5] but the reviews there are seldom worth anything. – I do hope your brother will be able to achieve his larger work, and thus have made his life thoroughly illustrious: and we must all be glad that he has set the world on this great new track meantime. It saves time, and will hasten the due acceptance of the one to come. . . .

To Fanny Wedgwood

March 13 / 60

Dear friend

I know now to be thankful for a long letter from a friend in the midst of the London season. I really am very grateful for yours; and I tell Maria I must write myself this time. – Your account of Eras: is better than I ventured to hope for. He is very good to think of sending me Buckle II.[6] I shall be delighted to have it. – It seemed to me, after I had written to him, that I ought to have said one thing more about C. D's [Charles Darwin's] book, for honesty's sake: and the notices I have seen have reminded me of this since, more than once. I rather regret that C. D. went out of his way two or three times (I think not more) to speak of 'the Creator' in the popular sense of the First Cause; and also *once* of the

'final cause' of certain cuckoo affairs. This latter is sure to be
misunderstood, in the face of all the rest of the book: and
the other gives occasion for people to ride off from the argument in a way which need not have been granted to them.
It is curious to see how those who would otherwise agree
with him turn away because his view is 'derived from', or
'based on', 'theology', while he admits critics, by this opening, who would otherwise have no business with the book
at all. It seems to me that having carried us up to the earliest
group of forms, or to the single primitive one, he and we
have nothing to do with how those few forms, or that one,
came there. His subject is the 'Origin of Species', and not
the origin of Organization; and it seems a needless mischief
to have opened the latter speculation at all. – There now! I
have delivered my mind. Not that it signifies a straw,
except, as I said, for truth's sake. – I do hope nothing will
prevent his accomplishing his larger work. In *that* case,
it may come true that the present one will be forgotten in
10 years: – hardly otherwise, I should think. . . .

NOTES

Harriet Martineau (1802–76) faced an appalling list of personal
problems over a lifetime. Her family's investments failed. Her
father, elder brother, and fiancé died before their time (her
fiancé's death was complicated by insanity); each death was a
wrenching experience. Thus forced to earn her living as an
author, she also had to confront, and triumph over, a large
number of health problems, including a growing deafness that
made normal conversations impossible. Despite these setbacks,
she was destined to become a prolific and popular writer. Over
half a century she produced an impressive array of works of
fiction, travel books, polemical studies, children's tales, educational prescriptions, a large-scale history of the Thirty Years'
Peace, and speculation on such matters as 'philosophical
necessity', the evolution of religions, mesmerism, clairvoyance

and moral obligations. Her controversial *Society in America* (1837), followed by *A Retrospect of Western Travel* (1838), expressed support for the Abolitionist cause at perhaps the period of its greatest unpopularity. She ruffled other feathers with her *Letters from Ireland* (1857), written shortly after the Famine years had devastated that country. The quality of her writing, always high, was marked from the beginning by an attractive sincerity and straightforwardness that the public appreciated.

She was also a hostess willing to entertain a variety of strong and even radical thinkers; she loved the interplay of personalities. Among those who attended her teas were Monckton Milnes, Sydney Smith, Bulwer, Thomas Malthus, Thomas Carlyle, and the brothers Erasmus and Charles Darwin. As her correspondence shows, she took a keen interest in the quarrels that immediately engulfed Darwin and his defenders after the publication of *The Origin of Species*. Her comments indicate that even his personal friends were disturbed by the radical implications of several passages in the text.

An additional note: After Charles married his first cousin, Emma Wedgwood, daughter of Josiah Wedgwood of Maer (and Charles's uncle), his elder brother, Erasmus, seemed to have fallen in love with Fanny Wedgwood, Emma's sister. The relationship was destined to remain platonic (she was married); it lasted a major part of his life. It is to Fanny that Harriet addressed one of the two letters excerpted here.

1. *The Origin of Species*.

2. Richard Owen (1804–92) reviewed the book in the *Edinburgh Review*, April 1860.

3. John Arthur Roebuck (1801–79) was an independent politician who held several Radical views that Harriet Martineau supported.

4. Thomas Henry Huxley wrote the review, printed in *The Times* on 26 December 1859.

5. The *Daily News* review also appeared on 26 December 1859.

6. Henry Thomas Buckle, *History of Civilization in England* (1861), vol. II.

'A Philosopher Comments on *The Origin of Species'*, in *An Autobiography*, by Herbert Spencer (New York: D. Appleton and Company, 1904), vol. II, pp. 57–8 [1859]

While these articles were in hand, the *Origin of Species* was published. That reading it gave me great satisfaction may be safely inferred. Whether there was any set-off to this great satisfaction, I cannot now say; for I have quite forgotten the ideas and feelings I had. Up to that time, or rather up to the time at which the papers by Mr Darwin and Mr Wallace, read before the Linnaean Society, had become known to me, I held that the sole cause of organic evolution is the inheritance of functionally-produced modifications. The *Origin of Species* made it clear to me that I was wrong; and that the larger part of the facts cannot be due to any such cause. Whether proof that what I had supposed to be the sole cause, could be at best but a part cause, gave me any annoyance, I cannot remember; nor can I remember whether I was vexed by the thought that in 1852 I had failed to carry further the idea then expressed, that among human beings the survival of those who are the select of their generation is a cause of development. But I doubt not that any such feelings, if they arose, were overwhelmed in the gratification I felt at seeing the theory of organic evolution justified. To have the theory of organic evolution justified, was of course to get further support for that theory of evolution at large with which, as we have seen, all my conceptions were bound up. Believing as I did, too, that right guidance, individual and social, depends on acceptance of evolutionary views of mind and of society, I was hopeful that its effects would presently be seen on educational methods, political opinions, and men's ideas about human life.

Obviously these hopes that benefical results would presently be wrought, were too sanguine. My confidence in the rationality of mankind was much greater then than it is now.

NOTE

Herbert Spencer (1820–1903), the best-known and most influential English philosopher of the second half of the nineteenth century, earned Darwin's attention and respect, even though Spencer's prolific publications used terms like 'The Law of Evolution', 'direct adaptation' and 'the Development Hypothesis' in ways that could not be reconciled with Darwin's findings. Yet Spencer understood that the sciences of his time were grappling with serious and ultimate issues. His emphasis on the continuing progress of the universe amounted to a large-hearted, though philosophically flawed, effort to work out a general formula for all knowledge garnered from increasingly specialized fields. Spencer's interest in evolution was well-established before 1859, but the continuing (and increasingly bitter) uproar caused by the publication of *The Origin of Species* persuaded Spencer that his own faith in the meliorism of a doctrine of organic evolution – to be demonstrated by improvements in 'educational methods, political opinions, and men's ideas about human life' – was 'too sanguine'.

John Stevens Henslow, 'Marriage Advice; Defending Darwin', in *Darwin and Henslow / The Growth of an Idea / Letters 1831–1860*, edited by Nora Barlow (Berkeley and Los Angeles: University of California Press, 1967), pp. 148–9, 205–7 [1838, 1860]

[*To: C. Darwin Esq., 36 Grt Marlborough St, London*
From: Professor J. S. Henslow]

Cambridge
16 Dec 1838

My dear Darwin, –

This day 15 years ago I entered on that state which, it rejoices my pericardium to think that you are about to enter – I have been remiss in not telling you so sooner, but I am sure you will not think me unmindful of your happiness from having added one more specimen of my carelessness to the many you have witnessed before – All I can wish you is, that you may experience as great content in the marriage state as I have done myself – & all the advice, which I need not give you, is, to remember that as you take your wife for better for worse, be careful to value the better & care nothing for the worse – Of course it is impossible for a lover to suppose for an instant that there can be any worse in the matter, but it is the prudent part of a husband, to provide that there shall be none – It is the neglect of this little particular which makes the marriage state of so many men worse than their single blessedness – if there is such a

thing – for it is now so long since I have enjoyed my double blessedness that I cannot fancy myself in my Bachelor days –
One piece of more serious advice I shall just venture to hint
at – that we do well to remember daily that our greatest
earthly blessings may be taken from us in a moment. So far
from this reflection annoying us & preventing our happiness from being as complete as earthly happiness can be –
I have my own experience to assure you that it encreases
happiness, & removes many an anxious mental care. But I
am afraid you will think I am writing a sermon. Only take it
in good part & believe that I most heartily wish you all joy
& prosperity – Is there a chance of your coming here this
Xmas – Mrs H. is anxious to know & bids me ask you –

Yr. ever affect^{ly}

J. S. Henslow

[*To: Sir Joseph Hooker*
From: Professor J. S. Henslow]

7 Downing Terrace, Cambridge

May 10, 1860

My dear Joseph,

I don't know whether you care to hear Phillips, who deliv-
ers the Rede Lecture in the Senate House next Tuesday at
2 p.m. It is understood that he means to attack the Darwinian hypothesis of Natural Selection.

Sedgwick's address last Monday was temperate enough
for his usual mode of attack, but strong enough to cast a *slur*
upon all who substitute hypotheses for strict inductions,
and as he expressed himself in regard to some of C.D.'s suggestions as *revolting* to his own sense of right and wrong,
and as Dr Clark, who followed him, spoke so unnecessarily
severely against Darwin's views, I got up, as Sedgwick had
alluded to me, and stuck up for Darwin as well as I could,
refusing to allow that he was guided by any but truthful
motives, and declaring that he himself believed he was

exalting and not debasing our views of a Creator, in attribut-
ing to him a power of imposing laws on the Organic World
by which to do his work, as effectually as his laws imposed
on the inorganic had done it in the Mineral Kingdom.

I believe I succeeded in diminishing, if not entirely removing, the chances of Darwin's being prejudged by many who
take their cue in such cases according to the views of those
they suppose may know something of the matter. Yesterday at my lectures I alluded to the subject, and showed how
frequently Naturalists were at fault in regarding as *species*,
forms which had (in some cases) been shown to be varieties,
and how legitimately Darwin had deduced his *inferences*
from positive experiment. Indeed I had on Monday replied
to a sneer (I don't mean from Sedgwick) at his pigeon results,
by declaring that the case *necessitated* an appeal to such
domestic experiments, and that this was the legitimate and
best way of proceeding for the detection of those laws
which we are endeavouring to discover.

I do not disguise my own opinion that Darwin has
pressed his hypothesis too far, but at the same time I assert
my belief that his Book is (as Owen described it to me) the
'Book of the Day'. I suspect the passages I marked in the
Edinburgh Review for the illumination of Sedgwick have produced an impression upon him to a certain extent. When I
had had my say, Sedgwick got up to explain, in a very few
words, his good opinion of Darwin, but that he wished it to
be understood that his chief attacks were directed against
Powell's[1] late Essay from which he quoted passages as
'from an Oxford Divine' that would astound Cambridge
men, as no doubt they do. He showed how greedily (if I
may so speak) Powell has adopted all Darwin has suggested,
and applied these suggestions (as if the whole were already
proved) to his own views.

I think I have given you a fair, though very hasty, view of
what happened and as I have just had a letter from Darwin,
and really have not a minute to spare for a reply this morning, perhaps you will send this to him, as he may like to
know, to some extent, what happened.

[*To: Professor Henslow*
No postmark]

Down Bromley Kent
May 14th 1860

My dear Henslow
I have been greatly interested by your letter to Hooker; & I must thank you from my heart for so generously defending me as far as you could against my powerful attackers. – Nothing which persons say hurts me for long, for I have entire conviction that I have not been influenced by bad feelings in the conclusions at which I have arrived. Nor have I published my conclusions without long deliberations & they were arrived at after far more study than the publick will ever know of or believe in. – I am certain to have erred in many points, but I do not believe so much as Sedgwick and Co. think. Is there any Abstract or Proceedings of the Cambridge Phil. Soc. published? If so & you could get me a copy I sh^d like to have one.
Believe me my dear Henslow I feel grateful to you on this occasion & for the multitude of kindnessess you have done me from my earliest days at Cambridge. –

Yours affectionately
C. Darwin

NOTES

The Reverend John Stevens Henslow (1796–1861), working with his close friend Adam Sedgwick, helped to found the Cambridge Philosophical Society. Triumphing over the time-consuming activities of ecclesiastical appointments (he was the Rector of Hitcham), he came into his own as Professor of Botany at Cambridge University (from 1827 until his death). He had been teaching mineralogy, but the vigour with which he undertook his new assignment, plus his spirited leading of walks and his hospitality at numerous soirées, made him a respected and well-loved figure. He attracted many of the most promising students; Darwin, a favourite, became a life-long friend, although he knew his views on evolution and nat-ural selection were not accepted by Henslow.

Henslow's recommendation of Darwin for the post of natur-alist aboard *HMS Beagle* (see pp. 96–97 above) – a crucial factor in Captain Fitz-Roy's evaluation of Darwin's credentials – was followed by the close attention and care Henslow bestowed on all the specimens that Darwin sent back to England. (Henslow had briefly considered the possibility of going himself.) Dar-win was later to tell Henslow's son-in-law, Sir Joseph Dalton Hooker, that he 'fully believe[d] a better man than Henslow never walked this earth'.

The correspondence between Henslow and Darwin is not bulky, but as Nora Barlow, the editor, pointed out, Henslow's letters, limited in number, were written at crucial moments in Darwin's life; they encouraged Darwin to continue his voyage aboard the *Beagle* when reasons for becoming pessimistic about its outcome were accumulating in Darwin's mind; they com-mented on his coming marriage; and they reaffirmed his faith in Darwin's honesty during the storm that grew after the pub-lication of *The Origin of Species*. On one occasion, he presided over a meeting of the natural history section of the British Association that considered its implications for botanists and other scientists. Darwin knew that his scientific hypotheses and conclusions were not always acceptable to Henslow, and indeed that some of them were barely understood by his for-mer mentor.

Henslow's career included the advancement of botanical education in various kinds of schools, the active furtherance of farmers' causes, and the establishment of the Ipswich Museum.

1.   The Revd Baden Powell (1796–1860) FRS, Savilian Pro-fessor of Geometry at Oxford. Author of *Essays on the Spirit of the Inductive Philosophy, the Unity of Worlds, the Philosophy of Cre-ation* (1855). See *Notes and Records of the Royal Society of London*, vol. 14, no. 1, p. 51, 'Some unpublished letters of Charles Dar-win', ed. Sir Gavin de Beer, FRS. [Nora Barlow's note.]

'The Great Debate', in *Life and Letters of Thomas Henry Huxley*, edited by Leonard Huxley (New York: D. Appleton and Company, 1901), vol. I, pp. 192–204 [1860]

The famous Oxford Meeting of 1860 was of no small importance in Huxley's career. It was not merely that he helped to
save a great cause from being stifled under misrepresentation and ridicule – that he helped to extort for it a fair hearing; it was now that he first made himself known in popular
estimation as a dangerous adversary in debate – a personal
force in the world of science which could not be neglected.
From this moment he entered the front fighting line in the
most exposed quarter of the field.

Most unluckily, no contemporary account of his own
exists of the encounter. Indeed, the same cause which prevented his writing home the story of the day's work nearly
led to his absence from the scene. It was known that Bishop
Wilberforce, whose first class in mathematics gave him, in
popular estimation, a right to treat on scientific matters,
intended to 'smash Darwin'; and Huxley, expecting that the
promised debate would be merely an appeal to prejudice in
a mixed audience, before which the scientific arguments of
the Bishop's opponents would be at the utmost disadvan-
tage, intended to leave Oxford that very morning and join
his wife at Hardwicke, near Reading, where she was staying
with her sister. But in a letter, quoted below, he tells how,
on the Friday afternoon, he chanced to meet Robert Chambers, the reputed author of the *Vestiges of Creation*, who
begged him 'not to desert them'. Accordingly he postponed

his departure; but seeing his wife next morning, had no occasion to write a letter.

Several accounts of the scene are already in existence: one in the *Life of Darwin* (vol. ii, p. 320), another in the 1892 *Life*, p. 236 *sq.*; a third that of *Lyell* (vol. ii, p. 335), the slight differences between them representing the difference between individual recollections of eye-witnesses. In addition to these I have been fortunate enough to secure further reminiscences from several other eye-witnesses.

Two papers in Section D, of no great importance in themselves, became historical as affording the opponents of Darwin their opportunity of making an attack upon his theory which should tell with the public. The first was on Thursday, June 28. Dr Daubeny[1] of Oxford made a communication to the Section, 'On the final causes of the sexuality of plants, with particular reference to Mr Darwin's work on the *Origin of Species*'.[2] Huxley was called upon to speak by the President, but tried to avoid a discussion, on the ground 'that a general audience, in which sentiment would unduly interfere with intellect, was not the public before which such a discussion should be carried on.'

This consideration, however, did not stop the discussion; it was continued by Owen. He said he 'wished to approach the subject in the spirit of the philosopher', and declared his 'conviction that there were facts by which the public could come to some conclusion with regard to the probabilities of the truth of Mr Darwin's theory'. As one of these facts, he stated that the brain of the gorilla 'presented more differences, as compared with the brain of man, than it did when compared with the brains of the very lowest and most problematical of the Quadrumana'.

Now this was the very point, as said above, upon which Huxley had made special investigations during the last two years, with precisely opposite results, such as, indeed, had been arrived at by previous investigators. Hereupon he replied, giving these assertions a 'direct and unqualified contradiction' and pledging himself to 'justify that unusual procedure elsewhere', – a pledge which was amply fulfilled in the pages of the *Natural History Review* for 1861.

Accordingly it was to him, thus marked out as the champion of the most debatable thesis of evolution, that, two days later, the Bishop addressed his sarcasms, only to meet with a withering retort. For on the Friday there was peace; but on the Saturday came a yet fiercer battle over the 'Origin', which loomed all the larger in the public eye, because it was not merely the contradiction of one anatomist by another, but the open clash between Science and the Church. It was, moreover, not a contest of bare fact or abstract assertion, but a combat of wit between two individuals, spiced with the personal element which appeals to one of the strongest instincts of every large audience.

It was the merest chance, as I have already said, that Huxley attended the meeting of the section that morning. Dr Draper[3] of New York was to read a paper on the 'Intellectual Development of Europe considered with reference to the views of Mr Darwin'. 'I can still hear', writes one who was present, 'the American accents of Dr Draper's opening address when he asked "Air we a fortuitous concourse of atoms?"' However, it was not to hear him, but the eloquence of the Bishop, that the members of the Association crowded in such numbers into the Lecture Room of the Museum, that this, the appointed meeting-place of the section, had to be abandoned for the long west room, since cut in two by a partition for the purposes of the library. It was not term time, nor were the general public admitted; nevertheless the room was crowded to suffocation long before the protagonists appeared on the scene, 700 persons or more managing to find places. The very windows by which the room was lighted down the length of its west side were packed with ladies, whose white handkerchiefs, waving and fluttering in the air at the end of the Bishop's speech, were an unforgettable factor in the acclamation of the crowd.

On the east side between the two doors was the platform. Professor Henslow, the President of the section, took his seat in the centre; upon his right was the Bishop, and beyond him again Dr Draper; on his extreme left was Mr Dingle, a clergyman from Lanchester, near Durham, with Sir J. Hooker and Sir J. Lubbock in front of him, and nearer the centre, Professor Beale of King's College, London, and Huxley.

The clergy, who shouted lustily for the Bishop, were massed in the middle of the room; behind them in the north-west corner a knot of undergraduates (one of these was T. H. Green,[4] who listened but took no part in the cheering) had gathered together beside Professor Brodie,[5] ready to lift their voices, poor minority though they were, for the opposite party. Close to them stood one of the few men among the audience already in Holy orders, who joined in – and indeed led – the cheers for the Darwinians.

So 'Dr Draper droned out his paper, turning first to the right hand and then to the left, of course bringing in a reference to the Origin of Species which set the ball rolling.'

An hour or more that paper lasted, and then discussion began. The President 'wisely announced *in limine* that none who had not valid arguments to bring forward on one side or the other would be allowed to address the meeting; a caution that proved necessary, for no fewer than four combatants had their utterances burked by him, because of their indulgence in vague declamation.'[6]

First spoke (writes Professor Farrar[7]) a layman from Brompton, who gave his name as being one of the Committee of the (newly formed) Economic section of the Association. He, in a stentorian voice, let off his theological venom. Then jumped up Richard Greswell[8] with a thin voice, saying much the same, but speaking as a scholar; but we did not merely want any theological discussion, so we shouted them down. Then a Mr Dingle got up and tried to show that Darwin would have done much better if he had taken him into consultation. He used the blackboard and began a mathematical demonstration on the question – 'Let this point A be man, and let that point B be the mawnkey.' He got no further; he was shouted down with cries of 'mawnkey'. None of these had spoken more than three minutes. It was when these were shouted down that Henslow said he must demand that the discussion should rest on *scientific* grounds only.

Then there were calls for the Bishop, but he rose and said he understood his friend Professor Beale had something to say first. Beale, who was an excellent histologist, spoke to the effect that the new theory ought to meet with fair discussion, but added, with great modesty, that he himself had not sufficient knowledge to discuss the subject adequately. Then the Bishop spoke the speech that you know, and the question about his mother being an ape, or his grandmother.

From the scientific point of view, the speech was of small value. It was evident from his mode of handling the subject that he had been 'crammed up to the throat', and knew nothing at first hand; he used no argument beyond those to be found in his *Quarterly* article, which appeared a few days later, and is now admitted to have been inspired by Owen. 'He ridiculed Darwin badly and Huxley savagely; but,' confesses one of his strongest opponents, 'all in such dulcet tones, so persuasive a manner, and in such well turned periods, that I who had been inclined to blame the President for allowing a discussion that could serve no scientific purpose, now forgave him from the bottom of my heart.'

The Bishop spoke thus 'for full half an hour with inimitable spirit, emptiness and unfairness'. 'In a light, scoffing tone, florid and fluent, he assured us there was nothing in the idea of evolution; rock-pigeons were what rock-pigeons had always been. Then, turning to his antagonist with a smiling insolence, he begged to know, was it through his grandfather or his grandmother that he claimed his descent from a monkey?'[10]

This was the fatal mistake of his speech. Huxley instantly grasped the tactical advantage which the descent to personalities gave him. He turned to Sir Benjamin Brodie, who was sitting beside him, and emphatically striking his hand upon his knee, exclaimed, 'The Lord hath delivered him into mine hands'. The bearing of the exclamation did not dawn upon Sir Benjamin until after Huxley had completed his 'forcible and eloquent' answer to the scientific part of the Bishop's argument, and proceeded to make his famous retort.[11]

On this [continues the writer in *Macmillan's Magazine*] Mr Huxley slowly and deliberately arose. A slight tall figure, stern and pale, very quiet and very grave,[12] he stood before us and spoke those tremendous words – words which no one seems sure of now, nor, I think, could remember just after they were spoken, for their meaning took away our breath, though it left us in no doubt as to what it was. He was not ashamed to have a monkey for his ancestor; but he would be ashamed to be connected with a man who used great gifts to obscure the truth. No one doubted his meaning, and the effect was tremendous. One lady fainted and had to be carried out; I, for one, jumped out of my seat.

The fullest and probably most accurate account of these concluding words is the following, from a letter of the late John Richard Green, then an undergraduate, to his friend, afterwards Professor Boyd Dawkins:[13] –

I asserted – and I repeat – that a man has no reason to be ashamed of having an ape for his grandfather. If there were an ancestor whom I should feel shame in recalling it would rather be a man – a man of restless and versatile intellect – who, not content with an equivocal[14] success in his own sphere of activity, plunges into scientific questions with which he has no real acquaintance, only to obscure them by an aimless rhetoric, and distract the attention of his hearers from the real point at issue by eloquent digressions and skilled appeals to religious prejudice.[15]

Further, Mr A. G. Vernon-Harcourt, F.R.S., Reader in Chemistry at the University of Oxford, writes to me:

The Bishop had rallied your father as to the descent from a monkey, asking as a sort of joke how recent this had been, whether it was his grandfather or further back. Your father, in replying on this point, first explained that the suggestion was of descent through thousands of generations from a common ancestor, and then went on to this effect – 'But if this question is treated, not as a matter for the calm investigation of science, but as a matter of sentiment, and if I am asked whether I would choose to be descended from the poor animal of low intelligence and stooping gait, who grins and chatters as we pass, or from a man, endowed with great ability and a splendid position, who should use these gifts' [here, as the point became clear, there was a great outburst of applause, which mostly drowned the end of the sentence] 'to discredit and crush humble seekers after truth, I hesitate what answer to make.'

No doubt your father's words were better than these, and they gained effect from his clear, deliberate utterance, but in outline and in *scale* this represents truly what was said.

After the commotion was over, 'some voices called for Hooker, and his name having been handed up, the President invited him to give his view of the theory from the Botanical side. This he did, demonstrating that the Bishop, by his own showing, had never grasped the principles of the "Origin", and that he was absolutely ignorant of the elements of botanical science. The Bishop made no reply, and the meeting broke up.'[16]

ACCOUNT OF THE OXFORD MEETING BY THE REV. W. H. FREEMANTLE (IN *Charles Darwin, his Life Told*, &c., 1892, P. 238)

The Bishop of Oxford attacked Darwin, at first playfully, but at last in grim earnest. It was known that the Bishop had written an article against Darwin in the last *Quarterly*

Review;[17] it was also rumoured that Professor Owen had been staying at Cuddesdon and had primed the Bishop, who was to act as mouthpiece to the great Palaeontologist, who did not himself dare to enter the lists. The Bishop, however, did not show himself master of the facts, and made one serious blunder. A fact which had been much dwelt on as confirmatory of Darwin's idea of variation, was that a sheep had been born shortly before in a flock in the North of England, having an addition of one to the vertebrae of the spine. The Bishop was declaring with rhetorical exaggeration that there was hardly any evidence on Darwin's side. 'What have they to bring forward?' he exclaimed. 'Some rumoured statement about a long-legged sheep'. But he passed on to banter: 'I should like to ask Professor Huxley, who is sitting by me, and is about to tear me to pieces when I have sat down, as to his belief in being descended from an ape. Is it on his grandfather's or his grandmother's side that the ape ancestry comes in?' And then taking a graver tone, he asserted, in a solemn peroration, that Darwin's views were contrary to the revelation of God in the Scriptures. Professor Huxley was unwilling to respond: but he was called for and spoke with his usual incisiveness and with some scorn: 'I am here only in the interests of science,' he said, 'and I have not heard anything which can prejudice the case of my august client.' Then after showing how little competent the Bishop was to enter upon the discussion, he touched on the question of Creation. 'You say that development drives out the Creator; but you assert that God made you: and yet you know that you yourself were originally a little piece of matter, no bigger than the end of this gold pencil-case.' Lastly as to the descent from a monkey, he said: 'I should feel it no shame to have risen from such an origin; but I should feel it a shame to have sprung from one who prostituted the gifts of culture and eloquence to the service of prejudice and of falsehood.'

Many others spoke. Mr Gresley, an old Oxford don,[18] pointed out that in human nature at least orderly development was not the necessary rule: Homer was the greatest of poets, but he lived 3000 years ago, and has not produced his like.

Admiral FitzRoy was present, and said he had often expostulated with his old comrade of the *Beagle* for entertaining views which were contradictory to the First Chapter of Genesis.

Sir John Lubbock[19] declared that many of the arguments by which the permanence of species was supported came to nothing, and instanced some wheat which was said to have come off an Egyptian mummy, and was sent to him to prove that wheat had not changed since the time of the Pharaohs; but which proved to be made of French chocolate. Sir Joseph (then Dr) Hooker spoke shortly, saying that he had found the hypothesis of Natural Selection so helpful in explaining the phenomena of his own subject of Botany, that he had been constrained to accept it. After a few words from Darwin's old friend, Professor Henslow, who occupied the chair, the meeting broke up, leaving the impression that those most capable of estimating the arguments of Darwin in detail saw their way to accept his conclusions.

Note. – Sir John Lubbock also insisted on the embryological evidence for evolution.

F. D.

T. H. HUXLEY TO FRANCIS DARWIN (*ibid.*)

June 27, 1891.

I should say that Freemantle's account is substantially correct, but that Green has the substance of my speech more accurately. However, I am certain I did not use the word, 'equivocal'.

The odd part of the business is, that I should not have been present except for Robert Chambers. I had heard of the Bishop's intention to utilize the occasion. I knew he had the reputation of being a first-class controversialist, and I was quite aware that if he played his cards properly, we should have little chance, with such an audience, of making an efficient defence. Moreover, I was very tired, and wanted to join my wife at her brother-in-law's country house near Reading, on the Saturday. On the Friday I met Chambers in the street, and in reply to some remark of his, about his going to the meeting, I said that I did not mean to attend it – did not see the good of giving up peace and quietness to be episcopally pounded. Chambers broke out into vehement remonstrances, and talked about my deserting them. So I said, 'Oh! if you are going to take it that way, I'll come and have my share of what is going on.'

So I came, and chanced to sit near old Sir Benjamin Brodie. The Bishop began his speech, and to my astonishment very soon showed that he was so ignorant that he did not know how to manage his own case. My spirits rose proportionately, and when he turned to me with his insolent question, I said to Sir Benjamin, in an undertone, 'The Lord hath delivered him into mine hands.'

That sagacious old gentleman stared at me as if I had lost my senses. But, in fact, the Bishop had justified the severest retort I could devise, and I made up my mind to let him have it. I was careful however, not to rise to reply, until the meeting called for me – then I let myself go.

In justice to the Bishop, I am bound to say he bore no malice, but was always courtesy itself when we occasionally met in after years. Hooker and I walked away from the meeting together, and I remember saying to him that this experience had changed my opinion as to the practical value of the art of public speaking, and that from that time forth I should carefully cultivate it, and try to leave off hating it. I did the former, but never quite succeeded in the latter effort.

I did not mean to trouble you with such a long scrawl when I began about this piece of ancient history. – Ever yours very faithfully,

T. H. HUXLEY.

In the evening there was a crowded conversazione in Dr Daubeny's rooms, and here, continues the writer in *Macmillan's*, 'everyone was eager to congratulate the hero of the day. I remember that some naive person wished "it could come over again"; Mr Huxley, with the look on his face of the victor who feels the cost of victory, put us aside saying, "Once in a lifetime is enough, if not too much."'
In a letter to me the same writer remarks –

> I gathered from Mr Huxley's look when I spoke to him at Dr Daubeny's that he was not quite satisfied to have been forced to take so personal a tone – it a little jarred upon his fine taste. But it was the Bishop who first struck the insolent note of personal attack.

Again, with, reference to the state of feeling at the meeting:

> I never saw such a display of fierce party spirit, the looks of bitter hatred which the audience bestowed – (I mean the majority) on us who were on your father's side – as we passed through the crowd we felt that we were expected to say 'how abominably the Bishop was treated' – or to be considered outcasts and detestable.

It was very different, however, at Dr Daubeny's, 'where,' says the writer of the account in *Darwin's Life*, 'the almost sole topic was the battle of the "Origin", and I was much struck with the fair and unprejudiced way in which the black coats and white cravats of Oxford discussed the question, and the frankness with which they offered their congratulations to the winners in the combat.'

The result of this encounter, though a check to the other side, cannot, of course, be represented as an immediate and complete triumph for evolutionary doctrine. This was precluded by the character and temper of the audience, most of whom were less capable of being convinced by the arguments than shocked by the boldness of the retort, although, being gentlefolk, as Professor Farrar remarks, they were disposed to admit on reflection that the Bishop had erred on the score of taste and good manners. Nevertheless, it was a noticeable feature of the occasion, Sir M. Foster tells me, that when Huxley rose he was received coldly, just a cheer of encouragement from his friends, the audience as a whole not joining in it. But as he made his points the applause grew and widened, until, when he sat down, the cheering was not very much less than that given to the Bishop. To that extent he carried an unwilling audience with him by the force of his speech. The debate on the ape question, however, was continued elsewhere during the next two years, and the evidence was completed by the unanswerable demonstrations of Sir W. H. Flower at the Cambridge meeting of the Association in 1862.

The importance of the Oxford meeting lay in the open resistance that was made to authority, at a moment when even a drawn battle was hardly less effectual than acknowledged victory. Instead of being crushed under ridicule, the new theories secured a hearing, all the wider, indeed, for the startling nature of their defence.

NOTES

Thomas Henry Huxley's posthumous reputation as a scientist, despite his own achievements and his rhetorically gifted writings and public appearances, is closely related to the life and fortunes of Charles Darwin. Huxley (1825–95), like Darwin, was a genuine autodidact, and had much less formal education than Darwin. Like Darwin and Hooker, he began his professional career by accepting an appointment – as surgeon rather than as naturalist; he had successfully completed medical studies at Charing Cross Hospital and the University of London – aboard the Admiralty vessel *Rattlesnake* (December 1846 to November 1850). He met Darwin in the early 1850s, and admired his work. After the publication of *The Origin of Species*, he would devote much of his writing and lecturing to a popularization and furtherance of its critical theses. Even so, he neither fully accepted Darwin's views on evolution nor considered them essential to his continuing research in palaeontology.

Huxley, independently of Darwin, earned all the honours that came his way. He was one of the great educators of Victorian England, at all levels but perhaps most influentially at the elementary level; the destructor of the concept of archetypes given wide currency by the writings of Baron Cuvier; a pioneer in the field of evolution, correcting (to cite one example of his courage in confronting and arguing down the Establishment on the basis of powerfully arranged, superior information) Richard Owen's speculative explanation of vertebrate skulls; the formulator of a new classification of the zoological regions of the world; and a serious investigator of philosophical and theological truths. (Carlyle, a major influence, was dismayed by the direction that Huxley's writings took.)

Leonard Huxley, his son, paid his father the traditional homage of a *Life and Letters*, and it is from this two-volume work that an extensive treatment of the widely noticed Oxford meeting of 1860 is excerpted. Darwin was not present when Bishop Wilberforce was bloodied by Huxley's eloquence as a debater (probably more bloodied than the Bishop realized at the time); but since the event became one of the most notorious in the raging controversy of 'Darwinism', its importance as a human drama is secure, and no review of Darwin's life can afford to pass it by. An incidental irony of the event, duly recorded here, is that, before the meeting took place, Huxley believed that the impending clash of views might take place in the wrong venue, and that a 'general audience' in attendance at the British Association might be swayed more by 'sentiment' than by the 'intellect'.

1. Charles Giles Bridle Daubeny (1795–1867), botanist, held successive professorships in chemistry, botany and rural economy at Oxford University. His condescending attitude toward *The Origin of Species* was widely shared by fellow scientists in the 1860s, but did not prevail: 'very liberal and candid, but scientifically weak'.

2. My best thanks are due to Mr F. Darwin for permission to quote his accounts of the meeting; other citations are from the *Athenaeum* reports of 14 July 1860. [Leonard Huxley's footnote; hereafter cited as LH.]

3. John William Draper's lecture lasted an hour, and as soon as it ended, Bishop Wilberforce rose to reply.

4. Thomas Hill Green (1836–82), a pupil of Benjamin Jowett, was to become a moral philosopher of international repute, and his exposition of the ideas of Kant and Hegel was influential on several generations of students at Oxford University.

5. Sir Benjamin Collins Brodie (1783–1862), a physician, sat next to Huxley. (Emma Darwin had consulted him about a health problem in 1853.)

6. *Life of Darwin, l.c.* [LH]

7. Canon of Durham. [LH]

8. Rev. Richard Greswell, BD, Tutor of Worcester College [LH]

9. *Life of Darwin, l.c.* [LH]

10. 'Reminiscences of a Grandmother', *Macmillan's Magazine*, October 1898. Professor Farrar thinks this version of what the Bishop said is slightly inaccurate. His impression is that the words actually used seemed at the moment flippant and unscientific rather than insolent, vulgar or personal. The Bishop, he writes, 'had been talking of the perpetuity of species of Birds; and then, denying *a fortiori* the derivation of the species Man from Ape, he rhetorically invoked the aid of *feeling*, and said, "If any one were to be willing to trace his descent through an ape as his *grandfather*, would he be willing to trace his descent similarly on the side of his *grandmother*?" His false humour was an attempt to arouse the antipathy about degrading *woman* to the quadrumana. Your father's reply showed there was vulgarity as well as folly in the Bishop's words; and the impression distinctly was, that the Bishop's party, as they left the room, felt abashed, and recognized that the Bishop had forgotten to behave like a perfect gentleman.' [LH]

11. The *Athenaeum* reports him as saying that Darwin's theory was an explanation of phenomena in Natural History, as the undulatory theory was of the phenomena of light. No one objected to that theory because an undulation of light had never

been arrested and measured. Darwin's theory was an explanation of facts, and his book was full of new facts, all bearing on his theory. Without asserting that every part of that theory had been confirmed, he maintained that it was the best explanation of the origin of species which had yet been offered. With regard to the psychological distinction between men and animals, man himself was once a monad – a mere atom, and nobody could say at what moment in the history of his development he became consciously intelligent. The question was not so much one of a transmutation or transition of species, as of the production of forms which became permanent.

Thus the short-legged sheep of America was not produced gradually, but originated in the birth of an original parent of the whole stock, which had been kept up by a rigid system of artificial selection. [LH]

12.   'Young, cool, quiet, scientific – scientific in fact and in treatment.' – J. R. Green. A certain piquancy must have been added to the situation by the superficial resemblance in feature between the two men, so different in temperament and expression. Indeed next day at Hardwicke, a friend came up to Mr Fanning and asked who his guest was, saying, 'Surely it is the son of the Bishop of Oxford.' [LH] John Richard Green (1837–83) was sent to Magdalen College school at the age of eight, and led a solitary but internally rich life in the undergraduate years that followed. He served as a clergyman, and, from 1868 on, as the Librarian at Lambeth Palace until his natural interest in history, as well as the need to guard his health more carefully, led him into a new career, that of historian. His best-known and hugely successful work, *A Short History of the English People* (1874), was followed by *The Making of England* (1882) and *The Conquest of England* (1883).

13.   Sir William Boyd Dawkins (1837–1929), geologist, was supported by Charles Darwin in an unsuccessful application for the Professorship of Geology at Cambridge University. He became, instead, the Professor of Geology at Owen's College, Manchester, from 1872 on.

14.   The writer in *Macmillan's* tells me: 'I cannot quite accept Mr J. R. Green's sentences as your father's, though I didn't doubt that they convey the sense; but then I think that only a

shorthand writer could reproduce Mr Huxley's singularly beautiful style – so simple and so incisive. The sentence given is much too "Green".' [LH]

15.   My father once told me that he did not remember using the word 'equivocal' in this speech. (See his letter below.) The late Professor Victor Carus had the same impression, which is corroborated by Professor Farrar. [LH] Dr Julius Victor Carus (1823–1903) was a German zoologist who took great pleasure in translating Darwin's books into his native language.

LH adds, in a separate footnote: As the late Henry Fawcett wrote in *Macmillan's Magazine*, 1860: – 'The retort was so justly deserved, and so inimitable in its manner, that no one who was present can ever forget the impression that it made.' [LH]

16.   *Life of Darwin, l.c.* [LH]

17.   It appeared in the ensuing number for July. [LH]

18.   William Gresley (1801–76), divine, helped to popularize the publications of the Tractarian movement; contributed to the Englishman's Library, 31 volumes (1831–46); and fought against scepticism wherever he encountered it.

19.   Sir John William Lubbock, Bart. (1803–65), was a banker, a barrister, and a distinguished pioneering astronomer. He lived at High Elms, near Downe, and hence was Darwin's neighbour.

'How Darwin Encouraged Galton', in *Memories of My Life*, by Francis Galton (London: Methuen, 1908), pp. 162–3, 169, 287–8, 290–1 [1855, 1869]

It was not long after my marriage that the character of a piece of work that lay before me was clearly perceived. It was ready to be taken in hand and most suitable to my powers. It was to aid others in the exploration of the then unknown parts of the world, especially of Africa, of whose total length as much had been seen by me in my two journeys as perhaps by any one else then living. Being placed on the Council of the Royal Geographical Society, I thoroughly utilized that position to fulfil my object. The ignorance of travellers in any one country of the arts of travel employed in others was great, and I tried to make a compendium of them all. Having easy access to every traveller of note in England, I read many books of travel, or rather skimmed them for the purpose. Amongst others, I turned over every page in Pinkerton's well-known series of large quarto volumes of the narratives of travellers.

The result was that sufficient material was gathered for the composition of a small book entitled the *Art of Travel* (Murray). It soon reached a second edition, and was afterwards rewritten and much enlarged to form a third edition, which was stereotyped, and even now continues to be sold. I also took considerable part in the first edition of the *Hints to Travellers* issued by the Geographical Society, which has long since quite outgrown its original form, all its chapters having been rewritten, each of them by experts. In its present shape it is a most trustworthy guide to travellers for such instrumental and other scientific work as they need to be acquainted with. The Anthropological 'Notes and Queries' are a similar and most useful compendium relating to that branch of science. I had some share in this, but by no means a large one.

I cannot resist quoting the following letter from my cousin Charles Darwin, the great naturalist, whose opinion as the author of the *Voyage of the Beagle* was naturally valued by me most highly. I had asked him for hints while engaged on the first edition of the *Art of Travel*, and sent him a copy of it, to which he now refers. This was four years before the publication of the *Origin of Species*:

DOWN, Jan. 10 [?1855]
MY DEAR GALTON, – I received your kind present yesterday. I always thought your idea of your Book a very good

one, and that you would do it capitally, and from what I have seen my forethought is, I am sure, *quite* justified. I hope that your volume will have a large sale, but what I fully expect is that it will have a long sale, and if you save from some disasters half a dozen explorers, I feel sure that you will think yourself well rewarded for all the trouble your volume must have cost you. – Believe me, my dear Galton, yours very truly,

C. DARWIN

• • •

Entries in old diaries recall many pleasant social meetings at home, whether dinners, breakfasts, or simple gatherings of friends, where there was generally some traveller or other lion of the day whom people were glad to meet. I made occasional excursions to visit Charles Darwin at Down, usually at luncheon-time, always with a sense of the utmost veneration as well as of the warmest affection, which his invariably hearty greeting greatly encouraged. I think his intellectual characteristic that struck me most forcibly was the aptness of his questionings; he got thereby very quickly to the bottom of what was in the mind of the person he conversed with, and to the value of it.

• • •

The publication in 1859 of the *Origin of Species* by Charles Darwin made a marked epoch in my own mental development, as it did in that of human thought generally. Its effect was to demolish a multitude of dogmatic barriers by a single stroke, and to arouse a spirit of rebellion against all ancient authorities whose positive and unauthenticated statements were contradicted by modern science.

I doubt, however, whether any instance has occurred in which the perversity of the educated classes in misunderstanding what they attempted to discuss was more painfully conspicuous. The meaning of the simple phrase 'Natural

Selection' was distorted in curiously ingenious ways, and Darwinism was attacked, both in the press and pulpit, by persons who were manifestly ignorant of what they talked about. This is a striking instance of the obstructions through which new ideas have to force their way. Plain facts are apprehended in a moment, but the introduction of a new Idea is quite another matter, for it requires an alteration in the attitude and balance of the mind which may be a very repugnant and even painful process. On my part, however, I felt little difficulty in connection with the *Origin of Species*, but devoured its contents and assimilated them as fast as they were devoured, a fact which perhaps may be ascribed to an hereditary bent of mind that both its illustrious author and myself have inherited from our common grandfather, Dr Erasmus Darwin.

I was encouraged by the new views to pursue many inquiries which had long interested me, and which clustered round the central topics of Heredity and the possible improvement of the Human Race. The current views on Heredity were at that time so vague and contradictory that it is difficult to summarize them both justly and briefly. Speaking generally, most authors agreed that all bodily and some mental qualities were inherited by brutes, but they refused to believe the same of man. Moreover, theologians made a sharp distinction between the body and mind of man, on purely dogmatic grounds. A few passages may undoubtedly be found in the works of eminent authors that are exceptions to this broard generalization, for the subject of human heredity had never been squarely faced, and opinions were lax. It seems hardly credible now that even the word heredity was then considered fanciful and unusual. I was chaffed by a cultured friend for adopting it from the French.

• • •

Hereditary Genius made its mark at the time, though subjected to much criticism, no small part of which was captious or shallow, and therefore unimportant. The verdict which I most eagerly waited for was that of Charles Darwin, whom I ranked far above all other authorities on such a matter. His letter, given below, made me most happy.

DOWN, BECKENHAM, KENT, S. E.

3rd December

MY DEAR GALTON, – I have only read about 50 pages of your book (to Judges), but I must exhale myself, else something will go wrong in my inside. I do not think I ever in all my life read anything more interesting and original – and how well and clearly you put every point! George,[1] who has finished the book, and who expressed himself in just the same terms, tells me that the earlier chapters are nothing in interest to the later ones! It will take me some time to get to these latter chapters, as it is read aloud to me by my wife, who is also much interested. You have made a convert of an opponent in one sense, for I have always maintained that, excepting fools, men did not differ much in intellect, only in zeal and hard work; and I still think this is an *eminently* important difference. I congratulate you on producing what I am convinced will prove a memorable work. I look forward with intense interest to each reading, but it sets me thinking so much that I find it very hard work; but that is wholly the fault of my brain and not of your beautifully clear style. – Yours most sincerely,

(Signed)   CH. DARWIN

The rejoinder that might be made to his remark about hard work, is that character, *including the aptitude for work*, is heritable like every other faculty.

NOTES

Francis Galton (1822–1911) was the grandson of the poet-naturalist Erasmus Darwin; Charles Darwin was his cousin.

The publication of *The Origin of Species* converted him into an enthusiastic researcher on more than one aspect of that book's findings. (He was proud to receive, in 1902, the Darwin Medal of the Royal Society.)

But Galton's brilliant career, which began with extensive travelling in remote (and sometimes dangerous) corners of the Middle East and Africa, developed in several other directions. Although sometimes described as an anthropologist, he was a significant contributor to research areas in psychology, statistical theory, geology, and family characteristics. He developed a system of fingerprinting that, with relatively minor modifications, was adopted worldwide. *Meteorographica, or Methods of Mapping the Weather* (1863) improved weather forecasting; his development of an anthropometric laboratory (1884–5) foreshadowed the establishment of the Biometric Laboratory at University College, London; his fascination with statistical methods led to several critical works on both the theory and practice of quantitative measurements. He is sometimes called the father of eugenic research; he wrote a much-discussed treatise, *Hereditary Genius* (1869), and he established a Chair of Eugenics at the University of London.

The first excerpt describes Galton's understandable pride in the commercial success of *The Art of Travel* (1855, with several revised editions thereafter), and his elation at receiving a commendatory letter from his cousin; other passages talk about his affection for Darwin, and his enthusiasm over the quality of the scientific work recorded in *The Origin of Species*.

1. George Howard Darwin (1845–1912), second son of Charles and Emma Darwin. Despite ill health and a rather ordinary academic record, he was to become a distinguished scientist. In 1883 he was appointed to the Chair of Astronomy and Experimental Philosophy at Cambridge University.

Francis Darwin, 'A Character Sketch by Darwin's Son', in *The Life and Letters of Charles Darwin*, edited by Francis Darwin (1887; reprinted New York: Basic Books, 1959), vol. I, pp. 88–107 [1860s–1882]

It is my wish in the present chapter to give some idea of my father's everyday life. It has seemed to me that I might carry out this object in the form of a rough sketch of a day's life at Down, interspersed with such recollections as are called up by the record. Many of these recollections, which have a meaning for those who knew my father, will seem colourless or trifling to strangers. Nevertheless, I give them in the hope that they may help to preserve that impression of his personality which remains on the minds of those who knew and loved him – an impression at once so vivid and so untranslatable into words.

Of his personal appearance (in these days of multiplied photographs) it is hardly necessary to say much. He was about six feet in height, but scarcely looked so tall, as he stooped a good deal; in later days he yielded to the stoop; but I can remember seeing him long ago swinging his arms back to open out his chest, and holding himself upright with a jerk. He gave one the idea that he had been active rather than strong; his shoulders were not broad for his height, though certainly not narrow. As a young man he must have had much endurance, for on one of the shore excursions from the *Beagle*, when all were suffering from want of water, he was one of the two who were better able than the rest to struggle on in search of it. As a boy he was active, and could jump a bar placed at the height of the 'Adam's apple' in his neck.

He walked with a swinging action, using a stick heavily shod with iron, which he struck loudly against the ground, producing as he went round the 'Sand-walk' at Down, a rhythmical click which is with all of us a very distinct remembrance. As he returned from the midday walk, often carrying the waterproof or cloak which had proved too hot, one could see that the swinging step was kept up by something of an effort. Indoors his step was often slow and laboured, and as he went upstairs in the afternoon he might be heard mounting the stairs with a heavy footfall, as if each step were an effort. When interested in his work he moved about quickly and easily enough, and often in the middle of dictating he went eagerly into the hall to get a pinch of snuff, leaving the study door open, and calling out the last words of his sentence as he went. Indoors he sometimes used an oak stick like a little alpenstock, and this was a sign that he felt giddiness.

In spite of his strength and activity, I think he must always have had a clumsiness of movement. He was naturally awkward with his hands, and was unable to draw at all well.[1] This he always regretted much, and he frequently urged the paramount necessity of a young naturalist making himself a good draughtsman.

He could dissect well under the simple microscope, but I think it was by dint of his great patience and carefulness. It was characteristic of him that he thought many little bits of skilful dissection something almost superhuman. He used to speak with admiration of the skill with which he saw Newport dissect a humble bee, getting out the nervous system with a few cuts of a fine pair of scissors, held, as my father used to show, with the elbow raised, and in an attitude which certainly would render great steadiness necessary. He used to consider cutting sections a great feat, and in the last year of his life, with wonderful energy, took the pains to learn to cut sections of roots and leaves. His hand was not steady enough to hold the object to be cut, and he employed a common microtome, in which the pith for holding the object was clamped, and the razor slid on a glass surface in making the sections. He used to laugh at himself, and at his own skill in section-cutting, at which he would say he was 'speechless with admiration'. On the other hand, he must have had accuracy of eye and power of co-ordinating his movements, since he was a good shot with a gun as a young man, and as a boy was skilful in throwing. He once killed a hare sitting in the flower-garden at Shrews-bury by throwing a marble at it, and, as a man, he once killed a cross-beak with a stone. He was so unhappy at hav-ing uselessly killed the cross-beak that he did not mention it for years, and then explained that he should never have thrown at it if he had not felt sure that his old skill had gone from him.

When walking he had a fidgetting movement with his fingers, which he has described in one of his books as the habit of an old man. When he sat still he often took hold of one wrist with the other hand; he sat with his legs crossed, and from being so thin they could be crossed very far, as may be seen in one of the photographs. He had his chair in the study and in the drawing-room raised so as to be much higher than ordinary chairs; this was done because sitting on a low or even an ordinary chair caused him some dis-comfort. We used to laugh at him for making his tall draw-ing-room chair still higher by putting footstools on it, and then neutralizing the result by resting his feet on another chair.

His beard was full and almost untrimmed, the hair being grey and white, fine rather than coarse, and wavy or friz-zled. His moustache was somewhat disfigured by being cut short and square across. He became very bald, having only a fringe of dark hair behind.

His face was ruddy in colour, and this perhaps made people think him less of an invalid than he was. He wrote to Dr Hooker (June 13, 1849), 'Every one tells me that I look quite blooming and beautiful; and most think I am sham-ming, but you have never been one of those.' And it must be remembered that at this time he was miserably ill, far worse than in later years. His eyes were bluish grey under deep overhanging brows, with thick bushy projecting eyebrows. His high forehead was much wrinkled, but otherwise his face was not much marked or lined. His expression showed no signs of the continual discomfort he suffered.

When he was excited with pleasant talk his whole manner was wonderfully bright and animated, and his face shared to the full in the general animation. His laugh was a free and sounding peal, like that of a man who gives himself sympathetically and with enjoyment to the person and the thing which have amused him. He often used some sort of gesture with his laugh, lifting up his hands or bringing one down with a slap. I think, generally speaking, he was given to gesture, and often used his hands in explaining anything (*e.g.* the fertilization of a flower) in a way that seemed rather an aid to himself than to the listener. He did this on occasions when most people would illustrate their explanations by means of a rough pencil sketch.

He wore dark clothes, of a loose and easy fit. Of late years he gave up the tall hat even in London, and wore a soft black one in winter, and a big straw hat in summer. His usual out-of-doors dress was the short cloak in which Elliot and Fry's photograph represents him leaning against the pillar of the verandah. Two peculiarities of his indoor dress were that he almost always wore a shawl over his shoulders, and that he had great loose cloth boots lined with fur which he could slip on over his indoor shoes. Like most delicate people he suffered from heat as well as from chilliness; it was as if he could not hit the balance between too hot and too cold; often a mental cause would make him too hot, so that he would take off his coat if anything went wrong in the course of his work.

He rose early, chiefly because he could not lie in bed, and I think he would have liked to get up earlier than he did. He took a short turn before breakfast, a habit which began when he went for the first time to a water-cure establishment. This habit he kept up till almost the end of his life.

I used, as a little boy, to like going out with him, and I have a vague sense of the red of the winter sunrise, and a recollection of the pleasant companionship, and a certain honour and glory in it. He used to delight me as a boy by telling me how, in still earlier walks, on dark winter mornings, he had once or twice met foxes trotting home at the dawning.

After breakfasting alone about 7.45, he went to work at once, considering the $1\frac{1}{2}$ hour between 8 and 9.30 one of his best working times. At 9.30 he came into the drawing-room for his letters – rejoicing if the post was a light one and being sometimes much worried if it was not. He would then hear any family letters read aloud as he lay on the sofa.

The reading aloud, which also included part of a novel, lasted till about half-past ten, when he went back to work till twelve or a quarter past. By this time he considered his day's work over, and would often say, in a satisfied voice, '*I've* done a good day's work.' He then went out of doors whether it was wet or fine; Polly, his white terrier, went with him in fair weather, but in rain she refused or might be seen hesitating in the verandah, with a mixed expression of disgust and shame at her own want of courage; generally, however, her conscience carried the day, and as soon as he was evidently gone she could not bear to stay behind.

My father was always fond of dogs, and as a young man had the power of stealing away the affections of his sister's pets; at Cambridge, he won the love of his cousin W. D. Fox's[2] dog, and this may perhaps have been the little beast which used to creep down inside his bed and sleep at the foot every night. My father had a surly dog, who was devoted to him, but unfriendly to every one else, and when he came back from the *Beagle* voyage, the dog remembered him, but in a curious way, which my father was fond of telling. He went into the yard and shouted in his old manner; the dog rushed out and set off with him on his walk, showing no more emotion or excitement than if the same thing had happened the day before, instead of five years ago. This story is made use of in the 'Descent of Man', 2nd Edit., p. 74.

In my memory there were only two dogs which had much connection with my father. One was a large black and white half-bred retriever, called Bob, to which we, as children, were much devoted. He was the dog of whom the story of the 'hot-house face' is told in the 'Expression of the Emotions'.

But the dog most closely associated with my father was the above-mentioned Polly, a rough, white fox-terrier. She was a sharp-witted, affectionate dog; when her master was going away on a journey, she always discovered the fact by the signs of packing going on in the study, and became low-spirited accordingly. She began, too, to be excited by seeing the study prepared for his return home. She was a cunning little creature, and used to tremble or put on an air of misery when my father passed, while she was waiting for dinner, just as if she knew that he would say (as he did often say) that 'she was famishing'. My father used to make her catch biscuits off her nose, and had an affectionate and mock-solemn way of explaining to her before-hand that she must 'be a very good girl'. She had a mark on her back where she had been burnt, and where the hair had re-grown red instead of white, and my father used to commend her for this tuft of hair as being in accordance with his theory of pangenesis; her father had been a red bull-terrier, thus the red hair appearing after the burn showed the presence of latent red gemmules. He was delightfully tender to Polly, and never showed any impatience at the attentions she required, such as to be let in at the door, or out at the verandah window, to bark at 'naughty people' a self-imposed duty she much enjoyed. She died, or rather had to be killed, a few days after his death.[3]

My father's midday walk generally began by a call at the greenhouse, where he looked at any germinating seeds or experimental plants which required a casual examination, but he hardly ever did any serious observing at this time. Then he went on for his constitutional – either round the 'Sand-walk', or outside his own grounds in the immediate neighbourhood of the house. The 'Sand-walk' was a narrow strip of land $1\frac{1}{2}$ acres in extent, with a gravel-walk round it. On one side of it was a broad old shaw with fair-sized oaks in it, which made a sheltered shady walk; the other side was separated from a neighbouring grass field by a low quickset hedge, over which you could look at what view there was, a quiet little valley losing itself in the upland country towards the edge of the Westerham hill, with hazel coppice and larch wood, the remnants of what was once a large wood, stretching away to the Westerham road. I have heard my father say that the charm of this simple little valley helped to make him settle at Down.

The Sand-walk was planted by my father with a variety of trees, such as hazel, alder, lime, hornbeam, birch, privet, and dogwood, and with a long line of hollies all down the exposed side. In earlier times he took a certain number of turns every day, and used to count them by means of a heap of flints, one of which he kicked out on the path each time he passed. Of late years I think he did not keep to any fixed number of turns, but took as many as he felt strength for. The Sand-walk was our play-ground as children, and here we continually saw my father as he walked round. He liked to see what we were doing, and was ever ready to sympathize in any fun that was going on. It is curious to think how, with regard to the Sand-walk in connection with my father, my earliest recollections coincide with my latest; it shows how unvarying his habits have been.

Sometimes when alone he stood still or walked stealthily to observe birds or beasts. It was on one of these occasions that some young squirrels ran up his back and legs, while their mother barked at them in an agony from the tree. He always found birds' nests even up to the last years of his life, and we, as children, considered that he had a special genius in this direction. In his quiet prowls he came across the less common birds, but I fancy he used to conceal it from me, as a little boy, because he observed the agony of mind which I endured at not having seen the siskin or gold-finch, or whatever it might have been. He used to tell us how, when he was creeping noiselessly along in the 'Big-Woods' he came upon a fox asleep in the daytime, which was so much astonished that it took a good stare at him before it ran off. A Spitz dog which accompanied him showed no sign of excitement at the fox, and he used to end the story by wondering how the dog could have been so faint-hearted.

Another favourite place was 'Orchis Bank', above the quiet Cudham valley, where fly- and musk-orchis grew among the junipers, and Cephalanthera and Neottia under the beech boughs; the little wood 'Hangrove', just above this, he was also fond of, and here I remember his collecting grasses, when he took a fancy to make out the names of all the common kinds. He was fond of quoting the saying of one of his little boys, who, having found a grass that his father had not seen before, had it laid by his own plate during dinner, remarking, 'I are an extraordinary grass-finder!'

My father much enjoyed wandering slowly in the garden with my mother or some of his children, or making one of a party, sitting out on a bench on the lawn; he generally sat, however, on the grass, and I remember him often lying under one of the big lime-trees, with his head on the green mound at its foot. In dry summer weather, when we often sat out, the big fly-wheel of the well was commonly heard spinning round, and so the sound became associated with those pleasant days. He used to like to watch us playing at lawn-tennis, and often knocked up a stray ball for us with the curved handle of his stick.

Though he took no personal share in the management of the garden, he had great delight in the beauty of flowers – for instance, in the mass of Azaleas which generally stood in the drawing-room. I think he sometimes fused together his admiration of the structure of a flower and of its intrinsic beauty; for instance, in the case of the big pendulous pink and white flowers of Dielytra. In the same way he had an affection, half-artistic, half-botanical, for the little blue Lobelia. In admiring flowers, he would often laugh at the dingy high-art colours, and contrast them with the bright tints of nature. I used to like to hear him admire the beauty of a flower; it was a kind of gratitude to the flower itself, and a personal love for its delicate form and colour. I seem to remember him gently touching a flower he delighted in; it was the same simple admiration that a child might have.

He could not help personifying natural things. This feeling came out in abuse as well as in praise – *e.g.* of some seedlings – 'The little beggars are doing just what I don't want them to.' He would speak in a half-provoked, half-admiring way of the ingenuity of a Mimosa leaf in screwing itself out of a basin of water in which he had tried to fix it. One might see the same spirit in his way of speaking of Sundew, earth-worms, &c.[4]

Within my memory, his only outdoor recreation, besides walking, was riding, which he took to on the recommendation of Dr Bence Jones, and we had the luck to find for him the easiest and quietest cob in the world, named 'Tommy'. He enjoyed these rides extremely, and devised a number of short rounds which brought him home in time for lunch. Our country is good for this purpose, owing to the number of small valleys which give a variety to what in a flat country would be a dull loop of road. He was not, I think, naturally fond of horses, nor had he a high opinion of their intelligence, and Tommy was often laughed at for the alarm he showed at passing and repassing the same heap of hedge-clippings as he went round the field. I think he used to feel surprised at himself, when he remembered how bold a rider he had been, and how utterly old age and bad health had taken away his nerve. He would say that riding prevented him thinking much more effectually than walking – that having to attend to the horse gave him occupation sufficient to prevent any really hard thinking. And the change of scene which it gave him was good for spirits and health.

Unluckily, Tommy one day fell heavily with him on Keston common. This, and an accident with another horse, upset his nerves, and he was advised to give up riding.

If I go beyond my own experience, and recall what I have heard him say of his love for sport, &c., I can think of a good deal, but much of it would be a repetition of what is contained in his 'Recollections'. At school he was fond of

bat-fives, and this was the only game at which he was skilful. He was fond of his gun as quite a boy, and became a
good shot; he used to tell how in South America he killed
twenty-three snipe in twenty-four shots. In telling the story
he was careful to add that he thought they were not quite
so wild as English snipe.

Luncheon at Down came after his midday walk; and here
I may say a word or two about his meals generally. He had a
boy-like love of sweets, unluckily for himself, since he was
constantly forbidden to take them. He was not particularly
successful in keeping the 'vows' as he called them, which he
made against eating sweets, and never considered them
binding unless he made them aloud.

He drank very little wine, but enjoyed, and was revived
by, the little he did drink. He had a horror of drinking, and
constantly warned his boys that any one might be led into
drinking too much. I remember, in my innocence as a small
boy, asking him if he had been ever tipsy; and he answered
very gravely that he was ashamed to say he had once drunk
too much at Cambridge. I was much impressed, so that I
know now the place where the question was asked.

After his lunch, he read the newspaper, lying on the sofa
in the drawing-room. I think the paper was the only non-
scientific matter which he read to himself. Everything else,
novels, travels, history, was read aloud to him. He took so
wide an interest in life, that there was much to occupy him
in newspapers, though he laughed at the wordiness of the
debates; reading them, I think, only in abstract. His interest
in politics was considerable, but his opinion on these matters was formed rather by the way than with any serious
amount of thought.

After he had read his paper, came his time for writing letters. These, as well as the MS. of his books, were written by
him as he sat in a huge horse-hair chair by the fire, his paper
supported on a board resting on the arms of the chair.
When he had many or long letters to write, he would dic-
tate them from a rough copy; these rough copies were written on the backs of manuscript or of proof-sheets, and were

almost illegible, sometimes even to himself. He made a rule of keeping *all* letters that he received; this was a habit which he learnt from his father, and which he said had been of great use to him.

He received many letters from foolish, unscrupulous people, and all of these received replies. He used to say that if he did not answer them, he had it on his conscience afterwards, and no doubt it was in great measure the courtesy with which he answered every one, which produced the universal and widespread sense of his kindness of nature, which was so evident on his death.

He was considerate to his correspondents in other and lesser things, for instance when dictating a letter to a foreigner he hardly ever failed to say to me, 'You'd better try and write well, as it's to a foreigner.' His letters were generally written on the assumption that they would be carelessly read; thus, when he was dictating, he was careful to tell me to make an important clause begin with an obvious paragraph 'to catch his eye', as he often said. How much he thought of the trouble he gave others by asking questions, will be well enough shown by his letters. It is difficult to say anything about the general tone of his letters, they will speak for themselves. The unvarying courtesy of them is very striking. I had a proof of this quality in the feeling with which Mr Hacon, his solicitor, regarded him. He had never seen my father, yet had a sincere feeling of friendship for him, and spoke especially of his letters as being such as a man seldom receives in the way of business: – 'Everything I did was right, and everything was profusely thanked for.'

He had a printed form to be used in replying to troublesome correspondents, but he hardly ever used it; I suppose he never found an occasion that seemed exactly suitable. I remember an occasion on which it might have been used with advantage. He received a letter from a stranger stating that the writer had undertaken to uphold Evolution at a debating society, and that being a busy young man, without time for reading, he wished to have a sketch of my father's views. Even this wonderful young man got a civil answer, though I think he did not get much material for his speech. His rule was to thank the donors of books, but not of pamphlets. He sometimes expressed surprise that so few people thanked him for his books which he gave away liberally; the letters that he did receive gave him much pleasure, because he habitually formed so humble an estimate of the value of all his works, that he was generally surprised at the interest which they excited.

In money and business matters he was remarkably careful and exact. He kept accounts with great care, classifying them, and balancing at the end of the year like a merchant. I remember the quick way in which he would reach out for his account-book to enter each cheque paid, as though he were in a hurry to get it entered before he had forgotten it. His father must have allowed him to believe that he would be poorer than he really was, for some of the difficulty experienced in finding a house in the country must have arisen from the modest sum he felt prepared to give. Yet he knew, of course, that he would be in easy circumstances, for in his 'Recollections' he mentions this as one of the reasons for his not having worked at medicine with so much zeal as he would have done if he had been obliged to gain his living.

He had a pet economy in paper, but it was rather a hobby than a real economy. All the blank sheets of letters received were kept in a portfolio to be used in making notes; it was his respect for paper that made him write so much on the backs of his old MS., and in this way, unfortunately, he destroyed large parts of the original MS. of his books. His feeling about paper extended to waste paper, and he objected, half in fun, to the careless custom of throwing a spill into the fire after it had been used for lighting a candle.

My father was wonderfully liberal and generous to all his children in the matter of money, and I have special cause to remember his kindness when I think of the way in which he paid some Cambridge debts of mine – making it almost seem a virtue in me to have told him of them. In his later years he had the kind and generous plan of dividing his surplus at the year's end among his children.

He had a great respect for pure business capacity, and often spoke with admiration of a relative who had doubled his fortune. And of himself would often say in fun that what he really *was* proud of was the money he had saved. He also felt satisfaction in the money he made by his books. His anxiety to save came in a great measure from his fears that his children would not have health enough to earn their own livings, a foreboding which fairly haunted him for many years. And I have a dim recollection of his saying, 'Thank God, you'll have bread and cheese,' when I was so young that I was rather inclined to take it literally.

When letters were finished, about three in the afternoon, he rested in his bedroom, lying on the sofa and smoking a cigarette, and listening to a novel or other book not scientific. He only smoked when resting, whereas snuff was a stimulant, and was taken during working hours. He took snuff for many years of his life, having learnt the habit at Edinburgh as a student. He had a nice silver snuff-box given him by Mrs Wedgwood of Maer, which he valued much – but he rarely carried it, because it tempted him to take too many pinches. In one of his early letters he speaks of having given up snuff for a month, and describes himself as feeling 'most lethargic, stupid, and melancholy'. Our former neighbour and clergyman, Mr Brodie Innes,[5] tells me that at one time my father made a resolve not to take snuff except away from home, 'a most satisfactory arrangement for me,' he adds, 'as I kept a box in my study to which there was access from the garden without summoning servants, and I had more frequently, than might have been otherwise the case, the privilege of a few minutes' conversation with my dear friend.' He generally took snuff from a jar on the hall table, because having to go this distance for a pinch was a slight check; the clink of the lid of the snuff jar was a very familiar sound. Sometimes when he was in the drawing-room, it would occur to him that the study fire must be burning low, and when some of us offered to see after it, it would turn out that he also wished to get a pinch of snuff.

Smoking he only took to permanently of late years, though on his Pampas rides he learned to smoke with the Gauchos, and I have heard him speak of the great comfort of a cup of *maté* and a cigarette when he halted after a long ride and was unable to get food for some time.

The reading aloud often sent him to sleep, and he used to regret losing parts of a novel, for my mother went steadily on lest the cessation of the sound might wake him. He came down at four o'clock to dress for his walk, and he was so regular that one might be quite certain it was within a few minutes of four when his descending steps were heard.

From about half-past four to half-past five he worked; then he came to the drawing-room, and was idle till it was time (about six) to go up for another rest with novel-reading and a cigarette.

Latterly he gave up late dinner, and had a simple tea at half-past seven (while we had dinner), with an egg or a small piece of meat. After dinner he never stayed in the room, and used to apologize by saying he was an old woman, who must be allowed to leave with the ladies. This was one of the many signs and results of his constant weakness and ill-health. Half an hour more or less conversation would make to him the difference of a sleepless night, and of the loss perhaps of half the next day's work.

After dinner he played backgammon with my mother, two games being played every night; for many years a score of the games which each won was kept, and in this score he took the greatest interest. He became extremely animated over these games, bitterly lamenting his bad luck and exploding with exaggerated mock-anger at my mother's good fortune.

After backgammon he read some scientific book to himself, either in the drawing-room, or, if much talking was going on, in the study.

In the evening, that is, after he had read as much as his strength would allow, and before the reading aloud began, he would often lie on the sofa and listen to my mother playing the piano. He had not a good ear, yet in spite of this he had a true love of fine music. He used to lament that his enjoyment of music had become dulled with age, yet within my recollection, his love of a good tune was strong. I never heard him hum more than one tune, the Welsh song 'Ar hyd y nos', which he went through correctly; he used also, I believe, to hum a little Otaheitan song. From his want of ear he was unable to recognize a tune when he heard it again, but he remained constant to what he liked, and would often say, when an old favourite was played, 'That's a fine thing; what is it?' He liked especially parts of Beethoven's symphonies, and bits of Handel. He made a little list of all the pieces which he especially liked among those which my mother played – giving in a few words the impression that each one made on him – but these notes are unfortunately lost. He was sensitive to differences in style, and enjoyed the late Mrs Vernon Lushington's[6] playing intensely, and in June 1881, when Hans Richter paid a visit at Down, he was roused to strong enthusiasm by his magnificent performance on the piano. He much enjoyed good singing, and was moved almost to tears by grand or pathetic songs. His niece Lady Farrer's[7] singing of Sullivan's 'Will he come' was a never-failing enjoyment to him. He was humble in the extreme about his own taste, and correspondingly pleased when he found that others agreed with him.

He became much tired in the evenings, especially of late years, when he left the drawing-room about ten, going to bed at half-past ten. His nights were generally bad, and he often lay awake or sat up in bed for hours, suffering much discomfort. He was troubled at night by the activity of his thoughts, and would become exhausted by his mind working at some problem which he would willingly have dismissed. At night, too, anything which had vexed or troubled him in the day would haunt him, and I think it was then that he suffered if he had not answered some troublesome person's letter.

The regular readings, which I have mentioned, continued for so many years, enabled him to get through a great deal of the lighter kinds of literature. He was extremely fond of novels, and I remember well the way in which he would anticipate the pleasure of having a novel read to him, as he lay down, or lighted his cigarette. He took a vivid interest both in plot and characters, and would on no account know beforehand, how a story finished; he considered looking at the end of a novel as a feminine vice. He could not enjoy any story with a tragical end, for this reason he did not keenly appreciate George Eliot, though he often spoke warmly in praise of 'Silas Marner'. Walter Scott, Miss Austen, and Mrs Gaskell, were read and re-read till they could be read no more. He had two or three books in hand at the same time – a novel and perhaps a biography and a book of travels. He did not often read out-of-the-way or old standard books, but generally kept to the books of the day obtained from a circulating library.

I do not think that his literary tastes and opinions were on a level with the rest of his mind. He himself, though he was clear as to what he thought good, considered that in matters of literary taste, he was quite outside the pale, and often spoke of what those within it liked or disliked, as if they formed a class to which he had no claim to belong.

In all matters of art he was inclined to laugh at professed critics, and say that their opinions were formed by fashion. Thus in painting, he would say how in his day every one admired masters who are now neglected. His love of pictures as a young man is almost a proof that he must have had an appreciation of a portrait as a work of art, not as a likeness. Yet he often talked laughingly of the small worth of portraits, and said that a photograph was worth any number of pictures, as if he were blind to the artistic quality in a painted portrait. But this was generally said in his attempts to persuade us to give up the idea of having his portrait painted, an operation very irksome to him.

This way of looking at himself as an ignoramus in all matters of art, was strengthened by the absence of pretence, which was part of his character. With regard to questions of taste, as well as to more serious things, he always had the courage of his opinions. I remember, however, an instance

that sounds like a contradiction to this: when he was looking at the Turners in Mr Ruskin's bedroom, he did not confess, as he did afterwards, that he could make out absolutely
nothing of what Mr Ruskin saw in them. But this little pretence was not for his own sake, but for the sake of courtesy
to his host. He was pleased and amused when subsequently
Mr Ruskin brought him some photographs of pictures (I
think Vandyke portraits), and courteously seemed to value
my father's opinion about them.

Much of his scientific reading was in German, and this
was a great labour to him; in reading a book after him, I was
often struck at seeing, from the pencil-marks made each
day where he left off, how little he could read at a time.
He used to call German the 'Verdammte', pronounced as if
in English. He was especially indignant with Germans,
because he was convinced that they could write simply if
they chose, and often praised Dr F. Hildebrand[8] for writing
German which was as clear as French. He sometimes gave a
German sentence to a friend, a patriotic German lady, and
used to laugh at her if she did not translate it fluently. He
himself learnt German simply by hammering away with a
dictionary; he would say that his only way was to read a
sentence a great many times over, and at last the meaning
occurred to him. When he began German long ago, he
boasted of the fact (as he used to tell) to Sir J. Hooker, who
replied, 'Ah, my dear fellow, that's nothing; I've begun it
many times.'

In spite of his want of grammar, he managed to get on
wonderfully with German, and the sentences that he failed
to make out were generally really difficult ones. He never
attempted to speak German correctly, but pronounced the
words as though they were English; and this made it not a
little difficult to help him, when he read out a German sentence and asked for a translation. He certainly had a bad ear
for vocal sounds, so that he found it impossible to perceive
small differences in pronunciation.

His wide interest in branches of science that were not
specially his own was remarkable. In the biological sciences

his doctrines make themselves felt so widely that there was something interesting to him in most departments of it. He read a good deal of many quite special works, and large parts of text books, such as Huxley's 'Invertebrate Anatomy', or such a book as Balfour's 'Embryology',[9] where the detail, at any rate, was not specially in his own line. And in the case of elaborate books of the monograph type, though he did not make a study of them, yet he felt the strongest admiration for them.

In the non-biological sciences he felt keen sympathy with work of which he could not really judge. For instance, he used to read nearly the whole of 'Nature', though so much of it deals with mathematics and physics. I have often heard him say that he got a kind of satisfaction in reading articles which (according to himself) he could not understand. I wish I could reproduce the manner in which he would laugh at himself for it.

It was remarkable, too, how he kept up his interest in subjects at which he had formerly worked. This was strikingly the case with geology. In one of his letters to Mr Judd he begs him to pay him a visit, saying that since Lyell's death he hardly ever gets a geological talk. His observations, made only a few years before his death, on the upright pebbles in the drift at Southampton, and discussed in a letter to Mr Geikie, afford another instance. Again, in the letters to Dr Dohrn, he shows how his interest in barnacles remained alive. I think it was all due to the vitality and persistence of his mind – a quality I have heard him speak of as if he felt that he was strongly gifted in that respect. Not that he used any such phrases as these about himself, but he would say that he had the power of keeping a subject or question more or less before him for a great many years. The extent to which he possessed this power appears when we consider the number of different problems which he solved, and the early period at which some of them began to occupy him.

It was a sure sign that he was not well when he was idle at any times other than his regular resting hours; for, as long as he remained moderately well, there was no break in the regularity of his life. Week-days and Sundays passed by alike, each with their stated intervals of work and rest. It is almost impossible, except for those who watched his daily life, to realize how essential to his well-being was the regular routine that I have sketched: and with what pain and difficulty anything beyond it was attempted. Any public appearance, even of the most modest kind, was an effort to him. In 1871 he went to the little village church for the wedding of his elder daughter,[10] but he could hardly bear the fatigue of being present through the short service. The same may be said of the few other occasions on which he was present at similar ceremonies.

I remember him many years ago at a christening; a memory which has remained with me, because to us children it seemed an extraordinary and abnormal occurrence. I remember his look most distinctly at his brother Erasmus's funeral, as he stood in the scattering of snow, wrapped in a long black funeral cloak, with a grave look of sad reverie.

When, after an interval of many years, he again attended a meeting of the Linnean Society, it was felt to be, and was in fact, a serious undertaking; one not to be determined on without much sinking of heart, and hardly to be carried into effect without paying a penalty of subsequent suffering. In the same way a breakfast-party at Sir James Paget's,[11] with some of the distinguished visitors to the Medical Congress (1881), was to him a severe exertion.

The early morning was the only time at which he could make any effort of the kind, with comparative impunity. Thus it came about that the visits he paid to his scientific friends in London were by preference made as early as ten in the morning. For the same reason he started on his journeys by the earliest possible train, and used to arrive at the houses of relatives in London when they were beginning their day.

He kept an accurate journal of the days on which he worked and those on which his ill health prevented him from working, so that it would be possible to tell how many were idle days in any given year. In this journal – a little yellow Letts's Diary, which lay open on his mantel-piece, piled on the diaries of previous years – he also entered the day on which he started for a holiday and that of his return.

The most frequent holidays were visits of a week to London, either to his brother's house (6 Queen Anne Street), or to his daughter's (4 Bryanston Street). He was generally persuaded by my mother to take these short holidays, when it became clear from the frequency of 'bad days', or from the swimming of his head, that he was being overworked. He went unwillingly, and tried to drive hard bargains, stipulating, for instance, that he should come home in five days instead of six. Even if he were leaving home for no more than a week, the packing had to be begun early on the previous day, and the chief part of it he would do himself. The discomfort of a journey to him was, at least latterly, chiefly in the anticipation, and in the miserable sinking feeling from which he suffered immediately before the start; even a fairly long journey, such as that to Coniston,[12] tired him wonderfully little, considering how much an invalid he was; and he certainly enjoyed it in an almost boyish way, and to a curious extent.

Although, as he has said, some of his aesthetic tastes had suffered a gradual decay, his love of scenery remained fresh and strong. Every walk at Coniston was a fresh delight, and he was never tired of praising the beauty of the broken hilly country at the head of the lake.

NOTES

Francis Darwin (1848–1925), third son of Charles and Emma Darwin, earned degrees in mathematics and the natural sciences tripos at Trinity College, Cambridge, before spending a brief period as physician-in-training at St George's Hospital, London. His real interest, research, asserted itself between 1874 and 1882, when he worked happily as secretary and botanical assistant to his father. After the death of his first wife, Amy, he lived at Down House with his parents. His book *The Movement of Plants* (1880) secured his reputation. After the death of his father in 1882, he lived in Cambridge, concentrating on water movement in plants, the botany classes that he taught, and the editing of Charles Darwin's letters. Of great use to several generations of students were two books put together from his lecture materials: *Practical Physiology of Plants* (1894) and *The Elements of Botany* (1895). Although he wrote important biographical articles on his brother George Darwin as well as Francis Galton and Joseph Hooker, among others, he, more than any other individual, established the pool of information that subsequent biographers draw from, with thanks, to this day. His character sketch of his father, forming part of the text of *Life and Letters of Charles Darwin* (three volumes, 1887), is lively, detailed, and in large part responsible for Charles Darwin's high reputation as a human being among both those who knew him and those who didn't. Francis Darwin also edited *More Letters of Charles Darwin* (two volumes, 1903) and *The Foundation of The Origin of Species* (1909).

1.   The figure representing the aggregated cell-contents in 'Insectivorous Plants' was drawn by him. [Note of Francis Darwin, hereafter referred to as FD.]

2.   William Darwin Fox (1805–80), Charles Darwin's second cousin, was the Vicar of Delamere. In 1827 Charles Darwin met him, and wrote in his Journal, 'and so commenced Entomology.' Several letters back and forth survive; a presentation copy of the first edition of *The Origin of Species* became a prized item in Fox's library.

3.   The basket in which she usually lay curled up near the fire in his study is faithfully represented in Mr Parson's drawing, 'The Study at Down', facing page 101. [FD]

4.   Cf. Leslie Stephen's 'Swift' (1882), p. 200, where Swift's inspection of the manners and customs of servants are compared to my father's observations on worms. 'The difference is', says Mr Stephen, 'that Darwin had none but kindly feelings for worms.' [FD]

5. The Reverend John Brodie Innes, Curate of Farnborough, Kent, and Vicar of Downe (1846–69), was a fast friend of Charles Darwin for more than thirty years, although the two men often disagreed.

6. Vernon Lushington (1832–1912) was County Court Judge for Surrey and Berkshire (1877–1900), and Judge of the High Court of the Admiralty. He married Henrietta Emma Darwin, the eldest daughter of Charles Darwin. His name is on the list of family friends invited to the funeral.

7. Emma Cecilia Farrer (1854–1946) married Sir Horace Darwin (1880), thus becoming Charles Darwin's daughter-in-law.

8. Dr Friedrich Hermann Gustav Hildebrand (1835–1915) was Professor of Botany at Freiburg, Germany. Charles Darwin, who read Dr Hildebrand's scientific papers in German, appreciated the clarity of his prose because he never attained the degree of fluency in the German language that he believed he needed in order to keep up with research developments in Germany.

9. Francis Maitland Balfour (1851–82), a sociable friend to Charles Darwin's sons, was Professor of Animal Morphology at Cambridge University. His book *A Treatise on Comparative Embryology* (two volumes, 1881) became a standard text. He died while mountain climbing, shortly after Charles Darwin's funeral, which he had attended as a 'personal friend'.

10. Henrietta Darwin (1843–1930) married Richard Buckley Litchfield (1831–1903), a lawyer and philanthropist, in 1871. She edited *Emma Darwin: A Century of Family Letters* (1904 and 1915).

11. Sir James Paget, Bart. (1814–89), was a surgeon at St Bartholomew's Hospital. Darwin, who described him as a 'charming' man, used some material that Paget sent him for the book *Queries about Expression* (1867).

12. Charles Darwin and his family went to Coniston, Lancashire, for a vacation (2–27 August 1879).

'The Deaths of Erasmus and Charles Darwin', in *Life and Letters of Sir Joseph Dalton Hooker*, edited by Leonard Huxley, vol. II, pp. 258–60, 466–8 [1881–2]

A little later [Hooker] exclaims to Darwin (18 June 1881):

> We have lost no end of friends this year, and it is difficult to resist the pessimist view of creation. When I look back, however, my beloved friend, to the days I have spent in intercourse with you and yours, that view takes wings to itself and flies away; it is a horrid world to be sure, but it could have been worse.

Two months later died Darwin's elder brother, Erasmus, a man whose intellectual gifts and great personal charm were, owing to ill health, only known within the circle of his immediate friends and relations.

To Charles Darwin

August 29, 1881.

I have just seen the announcement of your brother's death and must send you a few words of heartfelt sympathy. I have somehow come to think those the happiest who, like myself, lost an only brother when very young; it seems now as if they could then be best spared – a blunder no doubt – but we know better what we lose after having lived so long together as you and your brother have.

It was in your brother's house, near Park Lane, that I first became acquainted with you – and shall never forget his kind face and kinder welcome – that was nearly 40 years ago! – I well remember thinking him then quite an elderly man and yet I see he was then under forty.

But a heavier loss was soon to follow. On April 19, 1882, died Charles Darwin, the friend of forty years, in science the ally and inspirer, in personal affection and intimate sympathy the closest of his circle. Hooker's sorrow and weariness were broken in upon by the request for an obituary notice to appear in *Nature*. Happily he was spared this task to which he felt sadly unequal.

Kew: April 21, 1882.

DEAR HUXLEY, – Romanes,[1] after asking me to write the notice of Darwin for *Nature*, now telegraphs that you had, unknown to him, been asked by the Sub-editor to undertake it, and had accepted.

I am right glad of it, as I am utterly unhinged and unfit for work and am not feeling well in my praecordia, and have not been for some time – pray say nothing of this, but I sometimes fear I shall have to seek rest if I would not that it were found for me. Nothing but the feeling that I was shrinking from duty induced me to assent to Romanes's request.

If I can help you with any notice of Darwin's early life I will come over to you on Sunday.

Up to the time of his going to Cambridge, though he had flirted a little with Nat. Hist., he had no notion of pursuing it, and had devoted himself to fox-hunting and partridges.

I did not feel our loss yesterday, but to-day I am depressed terribly, and a touching letter from Mrs Darwin quite upset me.

I have heard nothing about the Abbey, though Spottiswoode promised to telegraph the answer to me. I have no fancy for the bitter taste of these ceremonials. – Ever, dear old boy, yours,

J. D. HOOKER.

Kew: April 24, 1882.

DEAR HUXLEY, – It is well indeed that I turned Darwin over to you – the only idea I had parallel to yours was a comparison with Faraday. I have sent your eloquent and most impressive éloge on to Keltie,[2] with a note to send proof to you.

You are right; it is too soon for any sort of biographical notice of life or works.

As for myself, I have had a ten days' bout of my Anginic pains, night and day, and am in a state of nervous worry, with Bentham failing fast (82) and pressing the Genera Plantarum on me, and no end of work in the Garden.

In short I have my warning note struck.

On the 26th Darwin was buried in Westminster Abbey. Hooker was one of the pall-bearers.

• • •

The great event of 1909 was the centenary of Darwin's birth. Of all the galaxy of notable men who saw the light in the *annus mirabilis* 1809, Darwin, least in the public eye, came to have the profoundest influence in the world, transcending, beyond all others, the limits of his own country and his own lifetime. It was fitting that this honour should be paid to his memory and his enduring inspiration by Cambridge, his old University, where, if Darwin himself had profited little save by Henslow's direction of his bent towards science, science had since sprung up lustily under the Darwinian impulse, and a strong personal link with his name was kept up by the active work in the University of his distinguished sons.

The proceedings extended over three days, the 22nd, 23rd, and 24th of June; 1500 invitations were sent out. The first evening there was a reception by the Chancellor, Lord Rayleigh,[3] in the Fitzwilliam Museum. Next morning, a presentation of addresses by delegates of Universities, Colleges, Academies, and Learned Societies, in the Senate House; in the afternoon, a garden party at Christ's College, in the evening, a banquet in the New Examination Hall, followed by a reception at Pembroke. On the Thursday, honorary degrees were conferred in the Senate House; the Rede Lecture delivered by Sir Archibald Geikie, P.R.S., and in the afternoon a garden party given by the members of the Darwin family in Trinity College. There was an exhibition also of portraits, books, and other objects of interest in connexion with Darwin, in the Old Library of Christ's, his own College.

It was a brilliant function, resplendent with the bright and many coloured academic robes of various distinctions from a hundred seats of learning in every quarter of the civilized world. Of the guests who represented science at large or some personal link with the Darwin tradition, over five hundred sat down to the great banquet, a polyglot assembly keyed to the highest appreciation, where the admirable interest of Mr Balfour's historic speech was only eclipsed by the sense of personal charm in Mr W. E. Darwin's reminiscences of his father. Simple, direct, instinct with the same rich, unassuming humanity that they affectionately depicted, his words seemed to reveal from a still living source the very qualities of his father. 'Now,' one who had met Darwin whispered to his neighbour, 'those who never saw him will be able to understand why Darwin was so much beloved by his friends.'

NOTES

For a note on Sir Joseph Dalton Hooker, see above, pp. 151–2.

1. George John Romanes (1848–94), biologist, was one of Darwin's closest friends. He taught at University College, London, and later at Oxford University.

2.   Sir John Scott Keltie (1840–1927) was an important figure in the history of the Royal Geographical Society, serving in successive appointments as its librarian, secretary, and Vice-president. He was a sub-editor of *Nature*, and later the editor of the *Geographical Journal* (1893–1915). He assumed responsibility for the geographical materials in the 10th edition of the *Encyclopaedia Britannica*; a particularly important book in his bibliography is *The Partition of Africa* (1893). [LH]

3.   John William Strutt, third Baron Rayleigh (1842–1919), was a good friend of Darwin's sons at Cambridge, and was on the list of close personal friends invited to Darwin's funeral. He won the Nobel Prize in 1910 for his work in physics; served as the Cavendish Professor of Experimental Physics (1879–84) and as Chancellor of the University (1908–19).

Thomas Henry Huxley, 'Charles Darwin', in *Nature* (27 April 1882); reprinted *Darwiniana / Essays* (New York: D. Appleton and Company, 1893), pp. 244–7 [1882]

Very few, even among those who have taken the keenest interest in the progress of the revolution in natural knowledge set afoot by the publication of 'The Origin of Species', and who have watched, not without astonishment, the rapid and complete change which has been effected both inside and outside the boundaries of the scientific world in the attitude of men's minds towards the doctrines which are expounded in that great work, can have been prepared for the extraordinary manifestation of affectionate regard for the man, and of profound reverence for the philosopher, which followed the announcement, on Thursday last, of the death of Mr Darwin.

Not only in these islands, where so many have felt the fascination of personal contact with an intellect which had no superior, and with a character which was even nobler than the intellect; but, in all parts of the civilized world, it would seem that those whose business it is to feel the pulse of nations and to know what interests the masses of mankind, were well aware that thousands of their readers would think the world the poorer for Darwin's death, and would dwell with eager interest upon every incident of his history. In France, in Germany, in Austro-Hungary, in Italy, in the United States, writers of all shades of opinion, for once unanimous, have paid a willing tribute to the worth of our great countryman, ignored in life by the official representatives of the kingdom, but laid in death among his peers in Westminster Abbey by the will of the intelligence of the nation.

It is not for us to allude to the sacred sorrows of the bereaved home at Down; but it is no secret that, outside that domestic group, there are many to whom Mr Darwin's death is a wholly irreparable loss. And this not merely because of his wonderfully genial, simple, and generous nature; his cheerful and animated conversation, and the infinite variety and accuracy of his information; but because the more one knew of him, the more he seemed the incorporated ideal of a man of science. Acute as were his reasoning powers, vast as was his knowledge, marvellous as was his tenacious industry, under physical difficulties which would have converted nine men out of ten into aimless invalids; it was not these qualities, great as they were, which impressed those who were admitted to his intimacy with involuntary veneration, but a certain intense and almost passionate honesty by which all his thoughts and actions were irradiated, as by a central fire.

It was this rarest and greatest of endowments which kept his vivid imagination and great speculative powers within due bounds; which compelled him to undertake the prodigious labours of original investigation and of reading, upon which his published works are based which made him accept criticisms and suggestions from anybody and every-body, not only without impatience, but with expressions of gratitude sometimes almost comically in excess of their value; which led him to allow neither himself nor others to be deceived by phrases, and to spare neither time nor pains in order to obtain clear and distinct ideas upon every topic with which he occupied himself.

One could not converse with Darwin without being reminded of Socrates. There was the same desire to find some one wiser than himself; the same belief in the sovereignty of reason; the same ready humour; the same sympathetic interest in all the ways and works of men. But instead of turning away from the problems of Nature as hopelessly insoluble, our modern philosopher devoted his whole life to attacking them in the spirit of Heraclitus and of Demo-critus, with results which are the substance of which their speculations were anticipatory shadows.

The due appreciation, or even enumeration, of these results is neither practicable nor desirable at this moment. There is a time for all things – a time for glorying in our ever-extend-ing conquests over the realm of Nature, and a time for mourning over the heroes who have led us to victory.

None have fought better, and none have been more fortu-nate, than Charles Darwin. He found a great truth trodden underfoot, reviled by bigots, and ridiculed by all the world; he lived long enough to see it, chiefly by his own efforts, irrefragably established in science, inseparably incorpo-rated with the common thoughts of men, and only hated and feared by those who would revile, but dare not. What shall a man desire more than this? Once more the image of Socrates rises unbidden, and the noble peroration of the 'Apology' rings in our ears as if it were Charles Darwin's farewell: –

'The hour of departure has arrived, and we go our ways – I to die and you to live. Which is the better, God only knows.'

Thomas Henry Huxley, 'The Darwin Memorial', in *Darwiniana / Essays* (New York: D. Appleton and Company, 1893), pp. 248–52 [1885]

Your Royal Highness, – It is now three years since the announcement of the death of our famous countryman, Charles Darwin, gave rise to a manifestation of public feeling, not only in these realms, but throughout the civilized world, which, if I mistake not, is without precedent in the modest annals of scientific biography.

The causes of this deep and wide outburst of emotion are not far to seek. We had lost one of these rare ministers and interpreters of Nature whose names mark epochs in the advance of natural knowledge. For, whatever be the ultimate verdict of posterity upon this or that opinion which Mr Darwin has propounded; whatever adumbrations or anticipations of his doctrines may be found in the writings of his predecessors; the broad fact remains that, since the publication and by reason of the publication, of 'The Origin of Species' the fundamental conceptions and the aims of the students of living Nature have been completely changed. From that work has sprung a great renewal, a true 'instauratio magna' of the zoological and botanical sciences.

But the impulse thus given to scientific thought rapidly spread beyond the ordinarily recognized limits of biology. Psychology, Ethics, Cosmology were stirred to their foundations, and the 'Origin of Species' proved itself to be the fixed point which the general doctrine of evolution needed in order to move the world. 'Darwinism', in one form or another, sometimes strangely distorted and mutilated, became an everyday topic of men's speech, the object of an abundance both of vituperation and of praise, more often than of serious study.

It is curious now to remember how largely, at first, the objectors predominated; but considering the usual fate of new views, it is still more curious to consider for how short a time the phase of vehement opposition lasted. Before twenty years had passed, not only had the importance of Mr Darwin's work been fully recognized, but the world had discerned the simple, earnest, generous character of the man, that shone through every page of his writings.

I imagine that reflections such as these swept through the minds alike of loving friends and of honourable antagonists when Mr Darwin died; and that they were at one in the desire to honour the memory of the man who, without fear and without reproach, had successfully fought the hardest intellectual battle of these days.

It was in satisfaction of these just and generous impulses that our great naturalist's remains were deposited in Westminster Abbey; and that, immediately afterwards, a public meeting, presided over by my lamented predecessor, Mr Spottiswoode, was held in the rooms of the Royal Society,[1] for the purpose of considering what further step should be taken towards the same end.

It was resolved to invite subscriptions, with the view of erecting a statue of Mr Darwin in some suitable locality; and to devote any surplus to the advancement of the biological sciences.

Contributions at once flowed in from Austria, Belgium, Brazil, Denmark, France, Germany, Holland, Italy, Norway, Portugal, Russia, Spain, Sweden, Switzerland, the United States, and the British Colonies, no less than from all parts of the three kingdoms; and they came from all classes of the community. To mention one interesting case, Sweden sent in 2296 subscriptions 'from all sorts of people', as the distinguished man of science who transmitted them wrote, 'from the bishop to the seamstress, and in sums from five pounds to two pence'.

The Executive Committee has thus been enabled to carry out the objects proposed. A 'Darwin Fund' has been created, which is to be held in trust by the Royal Society, and is to be employed in the promotion of biological research.

The execution of the statue was entrusted to Mr Boehm;[2] and I think that those who had the good fortune to know Mr Darwin personally will admire the power of artistic divination which has enabled the sculptor to place before us so very characteristic a likeness of one whom he had not seen.

It appeared to the Committee that, whether they regarded Mr Darwin's career or the requirements of a work of art, no site could be so appropriate as this great hall, and they applied to the Trustees of the British Museum for permission to erect it in its present position.

That permission was most cordially granted, and I am desired to tender the best thanks of the Committee to the Trustees for their willingness to accede to our wishes.

I also beg leave to offer the expression of our gratitude to your Royal Highness for kindly consenting to represent the Trustees to-day.

It only remains for me, your Royal Highness, my Lords and Gentlemen, Trustees of the British Museum, in the name of the Darwin Memorial Committee, to request you to accept this statue of Charles Darwin.

We do not make this request for the mere sake of perpetuating a memory; for so long as men occupy themselves with the pursuit of truth, the name of Darwin runs no more risk of oblivion than does that of Copernicus, or that of Harvey.

Nor, most assuredly, do we ask you to preserve the statue in its cynosural position in this entrance-hall of our National Museum of Natural History as evidence that Mr Darwin's views have received your official sanction; for science does not recognize such sanctions, and commits suicide when it adopts a creed.

No; we beg you to cherish this Memorial as a symbol by which, as generation after generation of students of Nature enter yonder door, they shall be reminded of the ideal according to which they must shape their lives, if they would turn to the best account the opportunities offered by the great institution under your charge.

NOTES

For a note on Thomas Henry Huxley, see above pp. 182–3.

Huxley read the text of this address (9 June 1885) in his capacity of President of the Royal Society. He acted on behalf of the Memorial Committee in handing over the statue of Charles Darwin to HRH the Prince of Wales, who accepted it as the representative of the Trustees of the British Museum.

1.   William Spottiswoode (1825–83), president of the London Mathematical Society (1870), president of the British Association (1878), and president of the Royal Society (1878).

2.   Sir Joseph Edgar Boehm, Bart. (1834–90), was also the sculptor of the statue of Queen Victoria at Windsor Castle, the monument of the Duke of Kent in St George's Chapel, the statue of Carlyle on the Thames Embankment (Chelsea), the much-praised sarcophagus of Dean Stanley in Westminster Abbey, and the equestrian statue of the Duke of Wellington in St Paul's Cathedral.

Index

The title of each selection and the pages covered by the selection are printed in boldface type. Brackets around a title indicate that the wording has been devised by the editor to identify the subject-matter of the selection.

'Ar hyd y nos' (song), 206
Abraxus grossulariata (magpie moth), 104
Abyssinia, 111
Academy, The (periodical), 108
Africa, 111–12
Agassiz, Louis, 123
Ainsworth, William Francis, 31–2
Allingham, William, ix
Alum, 157
Antarctic, 135
Appleton, Thomas, ix
Argyll, Duke of, 106
Argyllshire coast, 157
Assyria, 32
Athenaeum (club), 22, 157
Audubon, John James, 33
Austen, Jane, 207

Balfour, Francis Maitland, *A Treatise on Comparative Embryology*, 209
Balfour, Mr, 217
Bangor (Wales), 46
Barmouth (Wales), 38, 46
Bates, Henry Walter, 99, 101
Beagle Channel, 80
Beagle, H.M.S., 29, 37, 45, 55, 78–81, 88, 135, 179, 192, 202
Beale, Professor, 174
Beaufort, Francis, Sir, 67–9, 70, 78–9, 87
Beethoven, Ludwig van, 206
Bennett, Alfred William, 'The Theory of Natural Selection from a Mathematical Point of View', 105
Bentham, George, 216
Berwick, Lord, 35

Bible, the, 37
'Big-Woods', 198
Blue Book, 128
Bob (dog), 197
Boehm, Joseph Edgar, Sir, ix, 223
British Association for the Advancement of Science, 173
Brodie, Benjamin Collins, Sir, 174–6, 180
Bryanston Street, 211
Buckle, Henry Thomas, 163
Buckley, Arabella Burton, 113
Burchill, Mr, 70
Butler, Arthur Gray, 105
Butler, Revd Samuel, 9, 15–17, 30, 55
Byron, Lord, 28

Cambridge, 29, 201, 203, 215
Cambridge Chronicle (periodical), 124
Cambridge Philosophical Society, 124
Cambridge University, 37–45, 216
Cameron, Julia Margaret, ix
Capel Curig (Wales), 46
Carabidous (beetles), 42
Carlyle, Thomas, *The French Revolution*, 143; *Reminiscences*, 28
Case, George, Revd, 8–9, 14–15
Cephalanthera, 199
Chambers, Robert, 179–80; *Vestiges of the Natural History of Creation*, 98, 120–2, 128, 171
Charleville, Lady, xi
Chester, Mr, 70
Chile, 141
Christ's College (Cambridge), 217

Church of England, 37
Church Stretton road, 9
Cimex (insect), 9
Clark, Dr, 167
Clark, Andrew, 124
Coldstream, John, 32
College of Surgeons, 128
Compton Bay, 157
Coniston (Lancashire), 211
Conway (Wales), 46, 56
Copernicus, 223
Copley medal, 157
Cotton, Mr, 34
Cuddesdon, 178
Cudham valley, 199
Cumberland, 104
Cwm Idwal (Wales), 46, 57

Daily News (London), 161
Darwin, Anne Elizabeth
    (daughter), 91–5
Darwin, Caroline (sister), 6, 8,
    13, 35
Darwin, Catherine (sister),
    6, 7, 13
Darwin, Charles,
    ['*Autobiographical Fragment,
    An*'], **6–11**; ['*Death of Anne
    Elizabeth Darwin*'], **91–5**; *Descent
    of Man, The*, 106, 108, 109, 196;
    ['*Early Years, The*'], **12–46**;
    *Expressions of the Emotions in
    Man and Animals, The*, 100, 109;
    *Formation of Vegetable Mould,
    through the Action of Worms...
    The*, 110; ['*Joining Captain
    Fitz-Roy on "H.M.S. Beagle"*'],
    59–75; *Origin of Species, The*, 32,
    57, 100, 118–23, 128–31, 133,
    145–6, 148–50, 155, 157–8, 162,
    164–5, 172–3, 177, 181, 187–9,
    221; ['*Preliminary Notice*'], **1–4**;
    *Variation of Animals and Plants
    under Domestication, The*, 109;
    Voyage of the Beagle, 135, 187
Darwin, Emma (wife), 92–3,
    205, 215

Darwin, Erasmus (grandfather),
    x, 1–5; *Botanic Garden*, 27;
    'Zoönomia', 2–3, 32
Darwin, Erasmus (brother),
    9–10, 27–8, 30–1, 99, 160–1,
    210, 214–15
Darwin, F., Sir, 72
Darwin, Francis, Sir, 130, 179;
    ['*A Character Sketch by Darwin's
    Son*'], **192–211**
Darwin, George Howard (son),
    191
Darwin, Henrietta Emma
    (daughter), ix, 210
Darwin Memorial Committee,
    223
Darwin, Susan (sister), 67–9,
    70–3
Darwin, Susannah (mother),
    6–8, 10–11, 13
Darwin, Violetta (aunt), 3
Darwin, William Erasmus (son),
    217
Daubeny, Charles G. B., 172, 180
Dawes, Richard, Sir, 44
Dawkins, Boyd, Professor,
    40, 176
del Piombo, Sebastian, 40
Denton (Durham), 62
Derby, 3
Devil Island, 79
Dielytra, 199
Dingle, Mr, 173–5
Dobbin (horse), 10
Dohrn, Dr, 209
Down (in Bromley, Kent), 99,
    134, 139, 188, 192–212
Draper, John William,
    'Intellectual Development
    of Europe...', 173–4
Duncan, Andrew, 30–1
Dundee, 155
Durham, 173

Edinburgh, 204
Edinburgh Review (periodical),
    100, 168

Edinburgh University, 30–46, 136, 139
Edwardsia (plant), 141
Eliot, George, *Silas Marner*, 207
Elliot and Fry (photographers), 195
Ellon (Scotland), 156
Entomological Society, 102
Euclid, 38–9
Eyton, Thomas Campbell, 45

Farrer, Emma Cecilia, Lady (daughter-in-law), 206
Farrer, Thomas Henry, Sir, 85, 174, 182
Field, The (periodical), 102
Fitz-Roy, Captain Robert, 59–75, 179; ['*Darwin Helps Save the Lifeboats of H.M.S. Beagle*'], **78–81**
Fitzwilliam Museum (Cambridge University), 40, 217
Flower, William Henry, Sir, 182
Forbes, Edward, 140
Ford (boyhood-friend), 9
Foster, Sir M., 182
Fox, William Darwin, 41, 73–4
Fraser (periodical), 100
Freemantle, W. H., Revd, 177, 179
Freemasons, 18–19
French Revolution, 43
Freshwater, Isle of Wight, ix
Fucus loreus 33

Galapagos, 136
Galton, Francis, Sir, 28; *Art of Travel*, 187; *Hereditary Genius*, 189–90; *Hints to Travellers*, 187; ['*How Darwin Encouraged Galton*'], **186–190**
Garnett (schoolboy-friend), 15–16
Gaskell, Mrs, 207
Gauchos (South America), 205
Geikie, Archibald, Sir, 209, 217

Genera Plantarum, 216
Geological Society, 161
George, Henry, *Progress and Poverty*, 112–13
Gil Blas, 1
Glamis, 155
Glen Trieg, 157
'God Save the King' (hymn), 40
Gower Street (London), 91
Grafton, Duke of, 59
Graham, W., *Creed of Science*, 113
Grant, Asa, 144; *Manual*, 143
Greek Testament, 38
Green, Thomas Hill, 174, 179
Gregory, Mr, 128
Gresley, Mr, 178
Greswell, Richard, 174
Gulf Stream, 141
Gulf of Ancud (South America), 83

Hacon, Mr (solicitor), 202
Haile, Peter, 7
Handel, George Frederick, 206
'Hangrove', 199
Hardie, Dr, 32
Hardwicke (near Reading), 171
Harvey, Betty, 7
Harvey, William, 223
Hemipterous insects, 29
Henslow, John Stevens, 39, 40, 42–5, 56, 60–2, 64, 67, 69–70, 79, 124, 167, 173, 179; ['*Marriage Advice; Defending Darwin*'], 166–9
Henslow, John, Mrs, 167
Herbert, John Maurice, 40
Herschel, John F. W., Sir, *Introduction to the Study of Natural Philosophy*, 44
Hildebrand, Friedrich H. G., Dr, 208
Hill, Major, 35
Hitcham, 43
Homer, 17, 38

Hooker, Joseph Dalton, Six, ix–x, 110, 112, 149, 169, 173–4, 177, 197–80, 194, 208, 214; ['*Deaths of Erasmus and Charles Darwin, The*'], **214–17**; *Flora Antarctica*, 137, 139, 149; ['*Great Friendship, A*'], **133–54**

Hooker, Joseph, Mrs, 143

Hooker, William, Sir, 154–5

Hope, Thomas Charles, 30

Hopkins, Mr, 100

Horace, 17, 36

Horner, Leonard, 34

Hughes, Professor, 56

Humboldt, Friedrich H. A. von, Baron, *Personal Narrative of Travels...*, 44

Huxley, Thomas Henry, x, 158, 216; ['*Charles Darwin*'], **218–20**; ['*Darwin Memorial, The*'], **221–4**; ['*Great Debate, The*'], **171–82**; *Invertebrate Anatomy*, 209

Huxley, Thomas Henry, Mrs, 180

India, 145

Innes, Brodie, 204

Jameson, Robert, 33, 34

Jamieson, Thomas Francis, 156

Jenyns, Leonard, 43, 59

Jenyns, Soames, 43

Johnson, Samuel, 1

Jones, Bence, Dr, 200

Judd, Mr, 209

Kay-Shuttleworth, James Phillips, 33

Keltie, John Scott, Sir, 216

Kew Gardens, 141–3

King, Philip Parker, 74, 80

King's College Chapel (Cambridge University), 40

Kinnordy, 155

Kynaston, Edward, Sir, 71

Lamarck, Chevalier de, 32, 98, 158

Lamb, Charles, 27

Leighton, W. A., Revd, 14

Letts' Diary, 211

Linnean Library, 99, 143, 146, 210

'Little Go' (Cambridge exam), 38

Livingstone, David, 124–6

Llangollen (Wales), 46

lobelia (flower), 199

London, 195, 210

London Review (periodical), 131

Londonderry, 72

Longfellow, Henry Wadsworth, ix

Lubbock, John, Sir, 174, 179

Lushington, Vernon, Mrs, 206

Lyell, Charles, Sir, 22, 87, 99, 113, 126, 135–6, 147, 150, 158; *Antiquity of Man, The,* 149; ['*Sir Charles Lyell and "The Origin of Species"*'], **154–8**

Lyell, Charles, Mr and Mrs, 135–6

Macgillivray, William, 35

Mackenzie, Charles Frederick, Bishop, 125

Mackintosh, James, Sir, 36, 43

MacLeay, 130

Madagascar, 111–12

Madeira Islands, 69, 71

Maer, 16, 35–6, 46, 204

Malthus, Thomas Robert, *An Essay on the Principles of Population, An*, 97

Malvern Wells, 101

Martineau, Harriet, ['*Darwin as a Welcome Guest*'], **87**; ['*Family Friend comments on "The Origin of Species", A*'], **160–3**

Matthew, Patrick, 131

Medical Congress (1881), 210

Megatherium (sloth), 89

Mill, John Stuart, 143

Mill, William Hodge, *Analysis of the Exposition of the Creed by Pearson*, 37
Mivart, St John Jackson, *On the Genesis of Species*, 109
Monro III, Alexander, 31
Montevideo, 88
More, Alexander Goodman, 157
Murray (publisher), 118
Museum of Natural History (London), ix, 223
Musters, Charles, 72

National Portrait Gallery (London), ix
National Gallery (London), 40
Natural History Review (periodical), 173
Nature (periodical), 106, 111, 209, 215
Neottia, 199
Newcastle, 154
New Examination Hall (Cambridge University), 217
Newhaven, 32
New Zealand, 141
Newport, Mr, 193
Norway, 141
Notes and Queries (anthropological), 187

Oken, 123
Osmaston, 75
Otaheitan song, 206
Owen, Mr, 35
Owen, Richard, 88–9, 123, 128–31, 161, 172, 178
Owen, Richard, Revd ['*Owen Studies a Darwin Specimen*'], **88–9**; ['*Quarrel between Owen and Darwin*'], **128–31**
Oxford Meeting (1860) 171–82
Oxford University, 171, 172, 176

P. quadripunctatus (beetle), 41
Pachyderms, 123
Paget, James, Sir, 210

Paley, William, *Natural Theology*, 38–9
Panagaeus crux-major (beetle), 41
Park Lane, 215
Parkes, Samuel, *Chemical Catechism*, 30
Parkfield, 7
Parslow, Joseph, 143
Patagonia, 74
Peacock, George, 59–62, 67–8, 79
Pemberton, Mr, 25
Pembroke (Cambridge University), 217
Petty, William: *see* Sherburn
Phillips, John, 167
Philosophical Magazine (periodical), 46
Philosophical Society (Cambridge), 169
Philosophical Transactions, 3
Pictet, Mr, 100
Pigott, Miss, 25–6
Pinkerton's (travel books), 187
Pistyll Rhiadr (Wales), 10
Plas Edwards (Wales), 9, 29
Plato, 'Apology', 220
Plinian Society, 33
Pliny, 18
Plymouth, 73
Polly (dog), 197
Polyanthus (flower), 14
Pontobdella muricata, 33
Powell, Baden, Revd, 170
primrose (flower), 14
Punta Alta (South America), 83

Quadrumana, 172
Quarterly Review (periodical), 100, 109, 177–8
Queen Anne Street (London), 99, 211

Ramsay, Alexander, Sir, 44
Ramsay, Marmaduke, 44, 60, 62
Rayleigh, Lord (Chancellor), 217

Reading (Berkshire town), 180
Rede Lecture (Cambridge Senate
   House) 167, 217
Reynolds, Joshua, Sir, 40
Richter, Hans, 206
Rio Cancarana (South America),
   89
Rio de Janeiro, 69
Rio de Plata (South America), 74
Rio Negro (South America), 88
Roebuck, John Arthur, 161
Romanes, George John, 215
Royal Geographical Society, 187
Royal Medical Society, 33–4
Royal Society, 3, 44
Ruskin, John, 208

Salisbury Craigs, 154
San Carlos Harbour (South
   America), 83
'Sand-walk' (at Down House),
   192, 197–8
Sarawak, 96
Sarmiento, 79
Scalidotherium, 89
Scott, Walter, Sir, 28, 207
Sedgwick, Adam, x, 39, 45–6,
   55–7, 167–8; ['*Adam Sedgwick's
   Reaction to "The Origin of
   Species"'*], **118–26**; ['*The
   Walking Tour in North
   Wales'*], **55–7**
Senate House (Cambridge
   University), 217
Seward, Miss, xi
Shakespeare, William, 28
Sherburn (CD's misspelling of
   Shelburne), 21–2
Shrewsbury, 3, 15, 18, 26,
   31, 34, 55, 73, 194
Shropshire, 24, 26, 29, 34
Sikkim (India), 145
Snowdon (Wales) 35
Socrates, 220
South America, 60, 74, 201
Southampton, 209
Spectator (periodical), 123–4

Spencer, Herbert, ['*A Philosopher
   Comments on "The Origin of
   Species"'*], **164–5**
Spottiswoode, 215
St Mark's Crescent (London), 100
Stephens, John Maurice,
   *Illustrations of British
   Entomology*, 41
Strait of Magalhaens, 80
Straits of Magellan, 7
Sulivan, Bartholomew James, Sir
   ['*Impressions of Charles
   Darwin'*], **83–6**
Sulivan, Henry Norton, 85
Sullivan, Sir Arthur, 'Will he
   come' (song), 206

Tasmania, 137, 144
Tenerife (also Teneriffe),
   44–5, 71
Tennyson, Alfred, Lord, ix
Ternate, 96
Thomson, James, *Seasons*, 28
Thomson, William, Sir, 109, 138
Tierra del Fuego, 60, 79, 135, 139
Tommy (horse), 200
Toxodon (fossil), 88
Trafalgar Square, 134
Trinity College (Cambridge
   University), 217
Turner, Joseph M. W., 208
Turner, Mr, 44

Ullswater (Lake District), 113
University of College
   (London), 32

Vandyke, Sir Anthony, 208
Vernon-Garcourt, A. G., 176
Virgil, 17
volute shell, 45–6

Wales, Prince of, 28, 35, 45
Wallace, Alfred Russel,
   Darwinism, 105; ['*Developing
   Friendship with Darwin, A'*],
   96–144; *My Life*, 96–114;

Natural Selection and Tropical Nature, 99, 105, 114
Warren, Dr, 4
Waterloo, 8
Waterton, Charles, 33
Wedgwood, Catherine ['Kitty'], 8
Wedgwood, Josiah, 35–6, 63–6, 72
Wedgwood, Sarah Elizabeth (aunt), 8
Wedgwood, Sarah, Mrs, 204
Weir, John Jenner, 99, 105
Wells, William Charles, 131
Werner, Abraham Gottlob, 32
Wernerian Society, 33
Westerham, 198
Westminster Abbey, 215–16
Westminster Review (periodical), 102

Wheatstone, Charles, 3
Whewell, William, 43
White, Revd Gilbert, *Natural History and Antiquities of Selborne,* 30
Whitley, Revd C., 40
Wickham, John Clements, 86
Wilberforce, Samuel, Bishop, 171–82
Wilkins, George, 119
Wombwell's Menagerie, 89
Wonders of the World, 28–9
Wood, Mr, 70–1
Woodhouse, 45

Yarrell, William, 72

Zoological Society (London), 72
Zoologist (periodical), 100
Zygaena (insect), 10, 29